W9-ADM-632

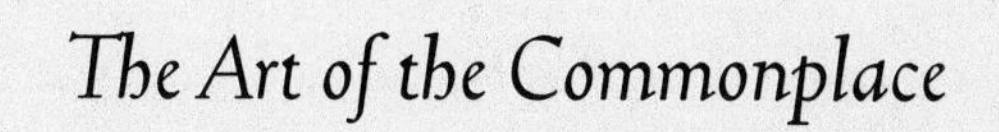
The Art of the Commonplace

Also by Wendell Berry

FICTION

The Discovery of Kentucky
Fidelity
Jayber Crow
The Memory of Old Jack
Nathan Coulter
A Place on Earth
Remembering
Three Short Novels
Two More Stories of the Port William Membership
Watch With Me
The Wild Birds
A World Lost

POETRY

The Broken Ground
Clearing
Collected Poems: 1957–1982
The Country of Marriage
Entries
Farming: A Hand Book
Findings
Openings
A Part
Sabbaths
Sayings and Doings
The Selected Poems of Wendell Berry (1998)
A Timbered Choir
Traveling at Home (with prose)
The Wheel

ESSAYS

Another Turn of the Crank
A Continuous Harmony
The Gift of Good Land
Harlan Hubbard: Life and Work
Life Is a Miracle
The Hidden Wound
Home Economics
Recollected Essays: 1954–1980
Sex, Economy, Freedom and Community
Standing by Words
The Unforeseen Wilderness
The Unsettling of America
What Are People For?

The Art of the COMMON-PLACE

The Agrarian Essays of Wendell Berry

EDITED AND INTRODUCED BY NORMAN WIRZBA

COUNTERPOINT WASHINGTON, D.C.

Library of Congress Cataloging-in-Publication Data
Berry, Wendell, 1934–
The art of the commonplace : agrarian essays of Wendell Berry / edited and introduced by Norman Wirzba.
p. cm.
ISBN 1-58243-146-9 (alk. paper)
1. Agriculture—United States. 2. Agriculture—Economic aspects—United States. 3. Agriculture—Social aspects—United States.
I. Wirzba, Norman. II. Title.
S441 .B47 2002
630'.973—dc21 20020007224

FIRST PRINTING

Jacket and text design by David Bullen Design

Printed in the United States of America on acid-free paper that meets the American National Standards Institute z39–48 Standard.

COUNTERPOINT
P.O. Box 65793
Washington, D.C. 20035–5793

Counterpoint is a member of the Perseus Books Group

10 9 8 7 6 5 4 3 2 1

Contents

The Challenge of Berry's Agrarian Vision

Novalis, the German romantic poet and philosopher, once remarked that all proper philosophizing is driven instinctually by the longing to be at home in the world, by the desire to bring to peace the restlessness that pervades much of human life. Wisdom, within this view, and therefore also happiness and well-being, rest on the ability to transpose the alienation and fragmentation that characterize life into a marriage of nature and spirit, or as William Wordsworth described it, a wedding between the human mind and "this goodly universe."

By most accounts, our efforts toward homecoming have ended in failure. Our unprecedented prosperity, rather than being founded in a convivial wholeness with the earth and with others, is predicated on the systematic exhaustion or destruction of life's sources—soil, water, and air—and the communities that inspire, define, and support our being. Our failure—as evidenced in flights to virtual worlds and the growing reliance on "life-enhancing" drugs, antidepressants, antacids, and stress management techniques—suggests a pervasive unwillingness or inability to make this world a home, to find in our places and communities, our bodies and our work, a joyful resting place. Perhaps even worse, we are training generations of children to see our anxiety-ridden ways as the norm.

Among contemporary writers, few have seen and described as clearly the causes and effects of our dis-ease as has Wendell Berry. In poetry, fiction, and essay he has not only characterized the destructive effects of our general homelessness but, more important, has also promoted a cogent and viable alternative in the form of agrarian thought and practice. Agrarianism, rather than being a quaint throwback to an impossible pastoral arcadia, is, in the hands and mind of Berry, a necessary and practical corrective to the waywardness of modern industrial culture. Agrarianism, in other words, promises a path toward wholeness with the earth, with each other, and with God, a path founded upon an insight into our proper place within the wider universe.

No doubt, an argument for agrarian life will sound old-fashioned. Most of us, after all, are urbanites, and so represent the culmination of a migratory trend that is unparalleled in human history. The exodus from country to city, now accomplished in "developed" countries and well on its way in the rest of the world, is seen by many as the necessary prerequisite for human freedom and progress. But given the enormity of implication and consequence that follows from a loss of intimate and practical connection with the earth (consider the general ignorance about our interdependence with other organisms), should we not at least pause to consider the cultural significance of the eclipse of agrarian sensibilities? It is Berry's contention that in abandoning what almost every culture has assumed and lived—that we are creatures intertwined in a common life with others—we bring harm to ourselves and the earth.

If we are to take agrarianism seriously, we must address two concerns. First, we must consider how the loss of agrarian ways is of broad cultural significance. And second, we must indicate how the recovery of agrarian principles and responsibilities can be a source for personal and cultural renewal.

* * *

> . . . Within things
> there is peace, and at the end
> of things. It is the mind
> turned away from the world
> that turns against it.
>
> (Wendell Berry, "Window Poems 19")

In the early essay "A Native Hill," Berry remarked that history testifies to an "intransigent destructiveness" within us that prompts us to place our well-being in contention with the well-being of others and the earth. We have assumed that we can know and pursue what is best for ourselves, all the while disregarding the needs of the communities, natural and human, that sustain us. The destructiveness of our much-touted progress has not always been widely apparent, especially in earlier times when the scale of the earth's bounty greatly outweighed our demands upon it. The destructiveness has now become unmistakably and unavoidably clear, however, particularly if we consider that many of the world's habitats and communities are in a condition of severe crisis as evident in soil erosion and toxification, water contamination and depletion, air pollution, deforestation, species extinction, destruction of rural communities, homogenization or decimation of indigenous cultures, and social cynicism and hopelessness. Though more of us than ever before live a life of luxury and ease, fewer of us can claim that our lives are permeated with peace and joy. The frantic, stressful striving going on all around us indicates that we are profoundly lost. We seem unable to ask with any seriousness or depth the question of what all our striving is ultimately for.

Where can we turn for help and direction? In the same essay Berry concluded that the source of help cannot come from within ourselves for "it is not from ourselves that we will learn to be better than we are." The path toward wholeness depends on our discovery and acknowledgment of, and then response to, a greater goodness that contextualizes us. Our fundamental mistake is that we have presumed to be the authors of ourselves and our destinies, and thus have forgotten or denied that we are part of "a great coauthorship in which we are all collaborating with God and with nature in the making of ourselves and one another." We can only become what we truly are by acknowledging that we do not exist by, from, and for ourselves. Our lives are always rooted in a natural and cultural community, so that to cut ourselves off from these roots, whether that be in the name of progress or human liberation, is to ensure the eventual withering and then death of life. Once we have forgotten or denied our biological kinship with the earth and its inhabitants, it is hardly an accident that so much of human spiritual life is premised on an escape from rather than an affirmation of this life.

The knowledge of our kinship with the earth could, in times past, be taken for granted, since the majority of people lived on and from the land. In the last hundred years this has all changed dramatically. We now find ourselves in societies where farmers are a statistically irrelevant minority (the number of prison inmates in this country exceeds the number of farmers). As farmland has been taken over by agribusiness, the intimate and concrete knowledge of our dependence on others, human and nonhuman, has been usurped by the industrial practice of human control and self-interest. The result is that many of the few who remain on the land have given up, whether by choice or by intense economic pressure, the agrarian ideals that have at various times punctuated and defined farming life.

Berry refers to agribusiness promoters as "the pornographers of farming," because their goal is maximum profit at whatever cost, and their method is rapacious exploitation. As Sir Albert Howard, a British agrarian much admired by Berry, once put it in *The Soil and Health:* "The using up of fertility is a transfer of past capital and of future possibilities to enrich a dishonest present: it is banditry pure and simple." The fact of the matter is that agribusiness practices cannot be sustained over the long term, because they depend on techniques that either produce a crop that often is equaled, even doubled, in weight by the soil lost to erosion, or kill the microbial, life-promoting soil structure with heavy pesticide use and soil compaction, or deplete and contaminate groundwater systems—all predicated on a cheap and limitless supply of fossil fuel. In our rush to maximize profits, or simply to perpetuate cash flow, we violate the sources of life on which we depend. Berry concludes: "The willingness to abuse other bodies is the willingness to abuse one's own. To damage the earth is to damage your children. To despise the ground is to despise its fruit; to despise the fruit is to despise its eaters. The wholeness of health is broken by despite" *(The Unsettling of America)*.

The effects of this violation are not limited to a few remaining farmers, but must necessarily be felt by culture as a whole, for as Berry insists, there is no such thing as a post-agricultural society—unless we learn to live without eating, drinking, and breathing.

Conventional wisdom would have us believe that the mass migration of farmers to urban life, sometimes voluntary and sometimes forced, is a good thing. After all, farmers enjoy no pensions, no paid vacations, no overtime pay, no unemployment compensation, no job security, and no specified hours of work; they have no time off they can call their own. According to Peter Drucker, the father of modern management theory, industrial work gave farmers their first opportunity to "better themselves substantially." As our society moves from an industrial to a knowledge-based economy, so the theory goes, the prospects for the improvement of workers' lives are getting increasingly better, since wages will be higher than for manufacturing jobs and workers will have "much greater opportunities." Gone is the day when we earn our keep with the sweat of our brow or with callused hands. This is the age of the acquisition, management, and distribution of knowledge.

What the purveyors and boosters of conventional wisdom often fail to ask, however, is whether the social and economic transformations they facilitate lead to an improper or inauthentic sense of human identity and vocation. A knowledge economy, like the industrial economy before it, demands that we see ourselves as specialists or careerists, trained to do a task of very limited scope and significance. Clearly, some forms of specialization are helpful, but we must not overlook the present reality in which specialization becomes an excuse to be ignorant of and nonaccountable for the broader contexts in which we work. We do not see or appreciate the biological and social fact that our lives and our responsibilities are complexly, yet harmoniously, intertwined with the lives of many others. The effect of careerism is thus to make ourselves frustratingly helpless and ignorant in regard to basic human skills—growing food, maintaining a home, caring for and educating children, promoting friendship and cooperation, facing illness and death—as well as financially dependent on other specialists who provide for us what we cannot provide for ourselves. Specialization also leads to the sense of our own isolation from the broader wholes of which we are a part. Isolated, we wonder how what we do matters, and perhaps more important, we shield ourselves from the harmful and the beneficial effects of what we do. Berry suggests that instead of measuring someone's intelligence in terms of the mastery of specialized information, we should instead judge it by "the good order or harmoniousness of his or her surroundings."

Consider how the difference between urban and agrarian life plays itself out with regard to the simple issue of waste. Most urbanites routinely place their garbage at the curb, "knowing" that it will be taken care of by sanitation specialists. In this act, however, we are largely ignorant as to where the garbage goes, how much of it there is, what effects it has on the habitats where it is finally deposited, and what consequences for others follow from the overall disposal process. Since our trained focus is on acquisition rather than disposal, we give no thought to the effects of some of our most basic acts.

In contrast, the agrarian thinks of waste in an entirely different manner, since his or her focus entails living with, and thus thinking about, the effects of what we do. Farmers know that dumps are not "sanitary" (they wouldn't want one on their land) and that they lead to conditions that are harmful in multiple, often unforeseen ways. And so rather than simply pass on the garbage we ourselves do not care to manage properly, the agrarian considers the flow and the consequences of waste production in all its particulars. The agrarian mind asks: Is the proposed product necessary? Can it be made to last indefinitely? How does this product's invention alter the quality of communal and natural life? Knowing the extreme difficulty of truly beneficial waste disposal, agrarians will want to limit from the start the practices that necessitate waste production, since the production of any waste represents a loss.

In the end, we or someone else will be left to live with our mess. Is it not naive and arrogant to think that there are no costs associated with our wasteful, consumptive ways, or that the costs will always be picked up by someone else?

Among the damaging effects of urban life is the loss of the practical sense of ourselves as embodied beings, as people who live under the necessary and practical constraints of biological life. Writing of the post–World War II era, an era marked by the mass migration of farmers to the city, Berry observed in *The Hidden Wound:*

> People had begun to live lives of a purely theoretical reality, daydreams based on the economics of success. It was as if they had risen off the earth into the purely hypothetical air of their ambition and greed. They were rushing around in the clouds, "getting somewhere," while their native ground, the only meaningful destination, if not the only possible one, lay far below them, abandoned and forgotten, colonized by machines.

Of course migration, though not without pain, is not necessarily destructive. But when migration is propelled by the fancies of a disembodied mind or desire—of a mind, for instance, that is focused on an abstract goal such as money (what constitutes "enough" cannot be determined or predicted)—then it is inevitable that "success" would require "the pragmatization of feeling," the forsaking of profitless pleasures and joys, the eclipse of meaningful leisure, the denigration of unuseful knowledge, and the denial of mystery. In Berry's view, we should not be surprised to find in this culture "marriage without love; sex without joy; drink without conviviality; birth, celebration, and death without adequate ceremony; faith without doubt or trial; belief without deeds; manners without generosity; 'good English' without exact speech, without honesty, without literacy."

One effect of the delusion of a purely intra-human life, of a life that is based and focused exclusively upon a humanly defined economy, is that we cease to be attentive to the biological, extra-human conditions that are indispensable for life's flourishing—to what Berry calls the "Great Economy" that surrounds and supports all our cultural efforts. Our lives become theoretical and abstract, focused on the hypothetically possible rather than the concrete and necessary present. Indeed the present is often sacrificed for the future. Because we do not attend to the present, and therefore fail to see what Gerard Manley Hopkins in his poem "God's Grandeur" described as "the dearest freshness deep down things," it is all but inevitable that our minds, whether out of blindness or hubris, will turn against the world. What we fail to see is that by turning against the world we finally turn against ourselves too.

The epigraph that begins Berry's *The Unsettling of America* reads: "Who so hath his minde on taking, hath it no more on what he hath taken." In this quotation from Montaigne, Berry alluded to how the transformation of cultures into consumptive, profit-driven organizations invariably leads to the exploitation of the earth and its inhabitants. The integrity of things, the sense that they are of value in themselves and in terms of the roles they play in a larger whole, is eclipsed by our desire to turn them to our benefit. Berry believes it is time for us to take full responsibility for the life-denying character of our histories and see in our damaged and exhausted landscapes the effects of a perverted desire. It is time for us to recognize that disembodied desire, desire cut off from the natural and social webs of interdependence, eventually leads to our own homelessness. We must, in other words, give up the age-old dream of the unencumbered human life, for as Berry insists in *The Unsettling of America:* "There is, in practice, no such thing as autonomy. Practically, there is only a distinction between responsible and irresponsible dependence."

* * *

Always, on their generation's breaking wave,
men think to be immortal in the world,
as though to leap from water and stand
in air were simple for a man. But the farmer
knows no work or act of his can keep him
here. He remains in what he serves
by vanishing in it, becoming what he never was.
He will not be immortal in words.
All his sentences serve an art of the commonplace,
to open the body of a woman or a field
to take him in. His words all turn
to leaves, answering the sun with mute
quick reflections . . .

(Wendell Berry, "The Farmer, Speaking of Monuments")

If the effect of an industrial, now increasingly knowledge-based economy is to turn us away from the world, the focus and inspiration of agrarian practice is the land. By "land" agrarians mean the life-giving sources of soil, water, and air, as well as the communities of organisms they support. Human life, no matter how much it may aspire to the realm of eternal, unchanging spirit, is thoroughly and necessarily embedded within the land. Our lives are literally constituted from out of the earth, since at the most basic level we ingest its gifts. And so the old scriptural view, that it is from the earth that we come and to the earth that we go upon death, is more than simply a figurative or poetic description. It goes to the heart of who we are and, by implication, what we are to become.

Needless to say, and with the exception of some indigenous and traditional societies, the goal of cultures has not been one of fidelity to the land. In fact, in many cases human progress has been measured in terms of our ability to combat and control the land, and so to establish human flourishing in opposition to the well-being of the earth. This is a temptation to which even farmers fall susceptible. But it is a temptation we must at all costs overcome, for what is at stake is the long-term happiness and health of ourselves and those we live with. Recognizing the general dis-ease of our own culture, we should now consider how agrarian insights and responsibilities serve as the point of departure for personal and cultural health.

A basic ecological principle is that nothing lives in isolation. As living beings, we are what we are only in terms of the many organisms and processes that together make life possible (even in death our bodies are involved in nature's life-promoting processes). What this means is that it is foolish to think that individual health and well-being can be maintained at the expense of the health of those on whom we depend. Why is it, then, that common conceptions of health postulate an individual freed from the effects of our association with others? Consider how disease is often characterized as the assault on the body by an alien agent, whereas physical and mental health are understood to be the liberation from a world of others known to be harmful or deadly. The presupposition here is that the world is an unfriendly place teeming with organisms that mean to do us harm. There is in this suspicious and fearful, albeit sanitary, view little trust in or gratitude for a world that is good and life-promoting.

According to agrarians, this view of health and disease, with its built-in antagonism, could only have come about in societies that no longer directly experience the beneficence of the earth. It rests on the false assumption that this earth is not a suitable home for us, that we cannot find here our joy and our peace (how much of the earth's present unsuitability is a reflection of our own activity?). It rejects the agrarian understanding summarized in *The Unsettling of America* that

> we and our country create one another, depend on one another, are literally part of one another; that our land passes in and out of our bodies just as our

bodies pass in and out of our land; that as we and our land are part of one another, so all who are living as neighbors here, human and plant and animal, are part of one another, and so cannot possibly flourish alone; that, therefore, our culture must be our response to our place, our culture and our place are images of each other and inseparable from each other, and so neither can be better than the other.

The fact of the matter is that countless societies have lived healthy and happy lives on the assumption that the earth can be our home. Moreover, and in direct contrast to modern cultures, these societies existed for centuries without destroying the sources on which their lives depended. It is as though life itself, whether conceived in human or nonhuman terms, has come to be understood by us as a burden, as something to be endured or suffered through, rather than as the effect of an unfathomable grace. We seem to have lost the ability to "work in the sun with hope, or sit at peace in the shade of any tree" ("The Morning's News").

Clearly what is needed is the recovery of the sense that our lives are necessarily and beneficially tied to the well-being of the earth, that the earth and its inhabitants form one vast "commonplace." In this recovery, two basic transformations must occur. On the one hand, we must learn to think very differently about the aims of life, of what counts as valuable knowledge or a noble pursuit. Given the countercultural thrust of this recovery, Berry terms our reeducation as a training in "a deathlier knowledge," a knowledge that not only takes full account of the grace-denying character of our own striving, but also acknowledges the fittingness of death in the processes of life. In his Sabbath poems Berry writes,

> The mind that comes to rest is tended
> In ways that it cannot intend:
> Is borne, preserved, and comprehended
> By what it cannot comprehend.
>
> Your Sabbath, Lord, thus keeps us by
> Your will, not ours. And it is fit
> Our only choice should be to die
> Into that rest, or out of it.

On the other hand, we must learn to reorganize our economic and social lives around the principle that health is an all-inclusive concept, a concept that involves soil, water, plants, animals, ecosystems, individuals, families, cities, and nations. The moment at which one of these elements is pitted against another is the moment of decline. Health, as Berry often says, is found in our harmonious membership with these elements. Morality, in turn, would represent our ability to dance within these patterns of interdependence.

In the poem "The Morning's News" Berry writes:

> I will purge my mind of the airy claims
> of church and state. I will serve the earth
> and not pretend my life could better serve.

The theme of "serving the earth" appears frequently in Berry's texts. It is easy to misunderstand, particularly since many of our assumptions about service come from an antagonistic worldview: to be a servant is to be servile, to be in a position where one can be taken advantage of. But if we assume, as agrarians say we must, that our lives are inextricably bound up with the lives of others, and that the context that informs life is gracious and merciful rather than vengeful or spiteful, then service takes on an entirely different meaning. Rather than being a demeaning activity, service becomes the ultimate act of life's promotion. Service is the "art of the commonplace," the art that willingly enters into life with others and the earth and seeks the flourishing of all. The labor of art, which here stands in contrast with the reductive, instrumental tendencies implicit in the desire to explain and control, seeks to expand our vision, and make it more faithful to the mystery and grace that comprehends and sustains us all.

The prerequisite for this art is the taming of our desire to the scale of the earth. What we must admit, however, is that our major cultural institutions do not prepare us very well for this task. In education, economics, politics, even religion, the scale of our striving is ourselves. This is a scale that easily lends itself to a combative mentality, because others are invariably perceived as a potential threat to my own aims. Service, or husbandry, as Berry sometimes calls it, works from the assumption that in another I find a real or potential ally, someone who already has assisted me (in ways often unnoticed by me), and can continue to assist me, in the goals of mutual growth and well-being. Rather than partaking in the anxious striving that characterizes a suspicious, fearful, or ungrateful life, service finds peace in the grace of the earth and the support of others. For good reason, Berry in the poem "Awake at Night" identifies "the last labor of the heart" as the ability "to learn to lie still, one with the earth again, and let the world go."

The model Berry often draws on to support this vision is the process of nature itself. Authentic art, art that serves the commonplace by uniting faithful perception and honest work, does not seek to exceed the bounds of propriety established by nature's rhythms. When considering a sycamore tree, Berry observes that this tree has lived entirely within its means. It has flourished by incorporating within itself what it is given—the nutrients of soil, water, and light, but also fence nails and lightning strikes. It has not tried to be more than it is. In "The Sycamore" Berry says, "I see that it stands in its place, and feeds upon it, and is fed upon, and is native, and maker." Its life, we might say, is the grateful and trusting response to a grace it neither intends nor knows. It simply takes for granted a comprehensive, though uncomprehended, beneficence and generosity out of and in which it lives and comes to rest.

In his Sabbath poems, collected in *A Timbered Choir,* Berry makes plain that the goal of human life, and therefore also its inspiration, must be to attain the peace and rest that marked the climax of creation. Our desire, our knowledge, and our work must finally end in the dance of joy and delight that mirrors the creator's pleasure in the goodness of creation. Berry's elevation of the significance of the Sabbath is not in the name of religious fastidiousness. Its focus is not simply the cessation from work one day of the week, but rather the transformation of all work in light of the goodness and interdependence of creation.

Be thankful and repay
Growth with good work and care.
Work done in gratitude
Kindly, and well, is prayer.

In articulating, however briefly, the vision that informs agrarian practice, a vision founded on the wholeness and health of the entire creation, we must not forget that agrarianism is above all a practice. For this reason Berry has focused much of his work on the careful examination of concrete cultural forms, offering critique and correction as he finds it necessary. In his view the vision and practices that currently inform our economic and social lives rest on assumptions that are not only false but destructive.

In promoting an agrarian culture, however, it is imperative that we not understand Berry to be saying that we must all become farmers. This is neither practically feasible nor desirable, since there is not enough land, nor does everyone have the appropriate temperament for farming. What is possible, however, is that people, urbanites included, adopt agrarian responsibilities and concerns. Just as we have adopted in our thought and practice the assumptions of an industrial mind-set without ourselves becoming industrialists—we are still teachers, health-care providers, builders, students, and so forth—so too can we integrate agrarian principles without ourselves becoming farmers.

What is involved? First, we must recognize that an agrarian transformation of contemporary culture will require the work of the imagination. We need to be able to envision a future that is markedly different from today's world, and be creative in the implementation of economic, political, religious, and educational reforms. No doubt this will be difficult, because the boosters of the industrial and knowledge economy proclaim the inevitability of where we are and where we are going. We should remember, however, that at the dawn of the industrial age, almost no one could have imagined a society in which farmers are statistically irrelevant. We should also consider that the boosters of the prevailing economic paradigm have the most to gain from its continuation, and that this economy is predicated on training us to be dissatisfied and ungrateful consumers.

Second, we must learn to resist those practices that further isolate us and turn us away from the earth. This can begin practically by cultivating a new relationship with food. Food, rather than simply being fuel, is the most concrete and intimate connection between ourselves and the earth that exists. We often have little sense of where food comes from or under what conditions it is produced. Educating ourselves or, better yet, participating in the production of our own food, as when we tend gardens, or getting to know and working with farmers more directly, will enable us to appreciate a variety of concerns—medical, economic, political, religious—from an agrarian point of view. It will force us to reconsider the meaning of health, and prompt us to see that the context of health extends beyond our own bodies to include rural communities and the land.

One of the unfortunate by-products of urbanization, one that has not gone unnoticed by contemporary urban planners, has been the isolation of city residents. For good reason, planners are trying to reinvigorate the idea of communal spaces, places where people can gather around common concerns. In this effort agrarians are helpful allies. Berry's own work has addressed at considerable length the obstacles to community formation. In many instances the primary obstacle is economic: corporate culture thrives on competition, the homogenization of desire, and the dream of ever-expanding markets. In this process, what is overlooked or ignored are the needs and limits of local traditions and places, the only meaningful place where communal life can occur. And so, contrary to the hype of globalization, Berry insists that the place of health and happiness must be grounded in the places where we now are. The hope for community finds its strongest support in developing local economies, for it is here, on the level of long-term, face-to-face exchange, that the work of cooperation and mutual help can find its most honest expression.

* * *

This collection represents a selection of some of Berry's essays that aim to move us beyond our present cultural malaise. They are not nostalgic: they conserve what is best from our past so that the problems of today can be adequately addressed. In his poem "At a Country Funeral," Berry spelled out his relation to the past precisely:

> . . . What we owe the future
> is not a new start, for we can only begin
> with what has happened. We owe the future
> the past, the long knowledge
> that is the potency of time to come.

The test of our lives is not the past or the future. It is the present. On this score it is plain to see that we are not doing well. Are we satisfied? That we live in an economy propelled by dissatisfaction and ingratitude clearly indicates that we are not. What is required, then, is that we honestly face our present circumstance. In "People, Land, and Community" Berry writes:

> One must stay to experience and study and understand the consequences—must understand them by living with them, and then correct them, if necessary, by longer living and more work. It won't do to correct mistakes made in one place by moving to another place, as has been the common fashion in America, or by adding on another place, as is the fashion in any sort of "growth economy.". . . It is the properly humbled mind in its proper place that sees truly, because—to give only one reason—it sees details.

What we need today is a renewed vision, a detailed insight into our own limits and possibilities that is informed by the knowledge of our place in the world. Berry's agrarian vision meets this need in a most remarkable way. It introduces us to the "art of the commonplace," and in so doing invites us to live a properly human life.

A Note to the Reader

Wendell Berry has published well over thirty books in the form of poetry, short story, novel, and essay. In selecting the following essays and excerpts, I have been guided by the desire to produce a synthetic account of his agrarian vision, all the while paying attention to how Berry's agrarian concerns might speak to a broad cultural audience. As should be clear from their reading, these essays touch on themes that are of perennial interest. They return us to the fundamental questions of human existence: Who are we? How does our life with others affect this self-understanding? What is a properly human desire? What are the limits and possibilities of communal life? How do we form an authentic culture? What are the conditions for a life of peace and joy?

Since much of Berry's writing refers to and is inspired by his farm home in Henry County, Kentucky, it is appropriate that Part I, "A Geobiography," introduce Berry's person and place to the reader. Part II considers how agrarian practice understands our contemporary cultural malaise. It describes the harmful and destructive effects that follow from the abandonment of our responsibilities to the earth and to each other. Part III provides the conceptual and practical foundations for an authentic life together. The resources for this foundation grow out of the sense of responsibility that accompanies our attentive care for each other and for the land. Part IV addresses the transformation in economy (understood in the broadest sense as "home management") that must follow from the acceptance of agrarian responsibilities. Part V turns to religion, to its critique and its transformation, for it is here that cultures have traditionally expressed their highest aspirations.

A brief editorial introduction orienting the reader to the essays will accompany each part of this collection.

Norman Wirzba

PART I

A Geobiography

Throughout his many texts Berry writes from the concrete and personal context of his own life and farm. "A Native Hill" provides a brief introduction to the person and place of Berry's work. Authentic and responsible thought, while not restricted to the local or regional, depends on the clarity and precision that comes from sustained attention to the particular.

> Pull down thy vanity, it is not man
> Made courage, or made order, or made grace,
> Pull down thy vanity, I say pull down.
> Learn of the green-world what can be thy place . . .
>
> Ezra Pound, *Canto LXXXI*

A Native Hill

I

The hill is not a hill in the usual sense. It has no "other side." It is an arm of Kentucky's central upland known as The Bluegrass; one can think of it as a ridge reaching out from that center, progressively cut and divided, made ever narrower by the valleys of the creeks that drain it. The town of Port Royal in Henry County stands on one of the last heights of this upland, the valleys of two creeks, Gullion's Branch and Cane Run, opening on either side of it into the valley of the Kentucky River. My house backs against the hill's foot where it descends from the town to the river. The river, whose waters have carved the hill and so descended from it, lies within a hundred steps of my door.

Within about four miles of Port Royal, on the upland and in the bottoms upriver, all my grandparents and great-grandparents lived and left such memories as their descendants have bothered to keep. Little enough has been remembered. The family's life here goes back to my mother's great-great-grandfather and to my father's great-grandfather, but of those earliest ones there are only a few vague word-of-mouth recollections. The only place antecedent to this place that has any immediacy to any of us is the town of Cashel in County Tipperary, Ireland, which one of my great-grandfathers left as a boy to spend the rest of his life in Port Royal. His name was James Mathews, and he was a shoemaker. So well did he fit his life into this place that he is remembered, even in the family, as having belonged here. The family's only real memories of Cashel are my own, coming from a short visit I made there five years ago.

And so, such history as my family has is the history of its life here. All that any of us may know of ourselves is to be known in relation to this place. And since I did most of my growing up here, and have had most of my most meaningful experiences here, the place and the history, for me, have been inseparable, and there is a sense in which my own life is inseparable from the history and the place. It is a complex inheritance, and I have been both enriched and bewildered by it.

* * *

I began my life as the old times and the last of the old-time people were dying out. The Depression and World War II delayed the mechanization of the farms here, and one of the first disciplines imposed on me was that of a teamster. Perhaps I first stood in the role of student before my father's father, who, halting a team in front of me, would demand to know which mule had the best head, which the best shoulder or rump, which was the lead mule, were they hitched right. And there came a time when I knew, and took a considerable pride in knowing. Having a boy's usual desire to play at what he sees men working at, I learned to harness and hitch and work a team. I felt distinguished by that, and took the same pride in it that other boys my age took in their knowledge of automobiles. I seem to have been born with an aptitude for a way of life that was doomed, although I did not understand that at the time. Free of any intuition of its doom, I delighted in it, and learned all I could about it.

That knowledge, and the men who gave it to me, influenced me deeply. It entered my imagination, and gave its substance and tone to my mind. It fashioned in me possibilities and limits, desires and frustrations, that I do not expect to live to the end of. And it is strange to think how barely in the nick of time it came to me. If I had been born five years later I would have begun in a different world, and would no doubt have become a different man.

Those five years made a critical difference in my life, and it is a historical difference. One of the results is that in my generation I am something of an anachronism. I am less a child of my time than the people of my age who grew up in the cities, or than the people who grew up here in my own place five years after I did. In my acceptance of twentieth-century realities there has had to be a certain deliberateness, whereas most of my contemporaries had them simply by being born to them.

* * *

In my teens, when I was away at school, I could comfort myself by recalling in intricate detail the fields I had worked and played in, and hunted over, and ridden through on horseback—and that were richly associated in my mind with people and with stories. I could recall even the casual locations of certain small rocks. I could recall the look of a hundred different kinds of daylight on all those places, the look of animals grazing over them, the postures and attitudes and movements of the men who worked in them, the quality of the grass and the crops that had grown on them. I had come to be aware of it as one is aware of one's body; it was present to me whether I thought of it or not.

When I have thought of the welfare of the earth, the problems of its health and preservation, the care of its life, I have had this place before me, the part representing the whole more vividly and accurately, making clearer and more pressing demands, than any *idea* of the whole. When I have thought of kindness or cruelty, weariness or exuberance, devotion or betrayal, carelessness or care, doggedness or awkwardness or grace, I have had in my mind's eye the men and women of this place, their faces and gestures and movements.

* * *

I have pondered a great deal over a conversation I took part in a number of years ago in one of the offices of New York University. I had lived away from Kentucky for several years—in California, in Europe, in New York City. And now I had decided to go back and take a teaching job at the University of Kentucky, giving up the position I then held on the New York University faculty. That day I had been summoned by one of my superiors at the university, whose intention, I had already learned, was to persuade me to stay on in New York "for my own good."

The decision to leave had cost me considerable difficulty and doubt and hard thought—for hadn't I achieved what had become one of the almost traditional goals of American writers? I had reached the greatest city in the nation; I had a good job; I was meeting other writers and talking with them and learning from them; I had reason to hope that I might take a still larger part in the literary life of that place. On the other hand, I knew I had not escaped Kentucky, and had never really wanted to. I was still writing about it, and had recognized that I would probably need to write about it for the rest of my life. Kentucky was my fate—not an altogether pleasant fate, though it had much that was pleasing in it, but one that I could not leave behind simply by going to another place, and that I therefore felt more and more obligated to meet directly and to understand. Perhaps even more important, I still had a deep love for the place I had been born in, and liked the idea of going back to be part of it again. And that, too, I felt obligated to try to understand. Why should I love one place so much more than any other? What could be the meaning or use of such love?

The elder of the faculty began the conversation by alluding to Thomas Wolfe, who once taught at the same institution. "Young man," he said, "don't you know you can't go home again?" And he went on to speak of the advantages, for a young writer, of living in New York among the writers and the editors and the publishers.

The conversation that followed was a persistence of politeness in the face of impossibility. I knew as well as Wolfe that there is a certain *metaphorical* sense in which you can't go home again—that is, the past is lost to the extent that it cannot be lived in again. I knew perfectly well that I could not return home and be a child, or recover the secure pleasures of childhood. But I knew also that as the sentence was spoken to me it bore a self-dramatizing sentimentality that was absurd. Home—the place, the countryside—was still there, still pretty much as I left it, and there was no reason I could not go back to it if I wanted to.

As for the literary world, I had ventured some distance into that, and liked it well enough. I knew that because I was a writer the literary world would always have an importance for me and would always attract my interest. But I never doubted that the world was more important to me than the literary world; and the world would always be most fully and clearly present to me in the place I was fated by birth to know better than any other.

And so I had already chosen according to the most intimate and necessary inclinations of my own life. But what keeps me thinking of that conversation is the feeling that it was a confrontation of two radically different minds, and that it was a confrontation with significant historical overtones.

I do not pretend to know all about the other man's mind, but it was clear that he wished to speak to me as a representative of the literary world—the world he assumed that I aspired to above all others. His argument was based on the belief that once one had attained the metropolis, the literary capital, the worth of one's origins was canceled out; there simply could be nothing *worth* going back to. What lay behind one had ceased to be a part of life, and had become "subject matter." And there was the belief, long honored among American intellectuals and artists and writers, that a place such as I came from could be returned to only at the price of intellectual death; cut off from the cultural springs of the metropolis, the American countryside is Circe and Mammon. Finally, there was the assumption that the life of the metropolis is *the* experience, the *modern* experience, and that the life of the rural towns, the farms, the wilderness places is not only irrelevant to our time, but archaic as well because unknown or unconsidered by the people who really matter—that is, the urban intellectuals.

I was to realize during the next few years how false and destructive and silly those ideas are. But even then I was aware that life outside the literary world was not without honorable precedent: if there was Wolfe, there was also Faulkner; if there was James, there was also Thoreau. But what I had in my mind that made the greatest difference was the knowledge of the few square miles in Kentucky that were mine by inheritance and by birth and by the intimacy the mind makes with the place it awakens in.

* * *

What finally freed me from these doubts and suspicions was the insistence in what was happening to me that, far from being bored and diminished and obscured to myself by my life here, I had grown more alive and more conscious than I had ever been.

I had made a significant change in my relation to the place: before, it had been mine by coincidence or accident; now it was mine by choice. My return, which at first had been hesitant and tentative, grew wholehearted and sure. I had come back to stay. I hoped to live here the rest of my life. And once that was settled I began to *see* the place with a new clarity and a new understanding and a new seriousness. Before coming back I had been willing to allow the possibility—which one of my friends insisted on—that I already knew this place as well as I ever would. But now I began to see the real abundance and richness of it. It is, I saw, inexhaustible in its history, in the details of its life, in its possibilities. I walked over it, looking, listening, smelling, touching, alive to it as never before. I listened to the talk of my kinsmen and neighbors as I never had done, alert to their knowledge of the place, and to the qualities and energies of their speech. I began more seriously than ever to learn the names of things—the wild plants and animals, the natural processes, the local places—and to articulate my observations and memories. My language increased and strengthened, and sent my mind into the place like a live root system. And so what has become the usual order of things reversed itself with me; my mind became the root of my life rather than its sublimation. I came to see myself as growing out of the earth like the other native animals and plants. I saw my body and my daily motions as brief coherences and articulations of the energy of the place, which would fall back into it like leaves in the autumn.

* * *

In this awakening there has been a good deal of pain. When I lived in other places I looked on their evils with the curious eye of a traveler; I was not responsible for them; it cost me nothing to be a critic, for I had not been there long, and I did not feel that I would stay. But here, now that I am both native and citizen, there is no immunity to what is wrong. It is impossible to escape the sense that I am involved in history. What I am has been to a considerable extent determined by what my forebears were, by how they chose to treat this place while they lived in it; the lives of most of them diminished it, and limited its possibilities, and narrowed its future. And every day I am confronted by the question of what inheritance I will leave. What do I have that I am using up? For it has been our history that each generation in this place has been less welcome to it than the last. There has been less here for them. At each arrival there has been less fertility in the soil, and a larger inheritance of destructive precedent and shameful history.

I am forever being crept up on and newly startled by the realization that my people established themselves here by killing or driving out the original possessors, by the awareness that people were once bought and sold here by my people, by the sense of the violence they have done to their own kind and to each other and to the earth, by the evidence of their persistent failure to serve either the place or their own community in it. I am forced, against all my hopes and inclinations, to regard the history of my people here as the progress of the doom of what I value most in the world: the life and health of the earth, the peacefulness of human communities and households.

And so here, in the place I love more than any other and where I have chosen among all other places to live my life, I am more painfully divided within myself than I could be in any other place.

* * *

I know of no better key to what is adverse in our heritage in this place than the account of "The Battle of the Fire-Brands," quoted in Collins's *History of Kentucky* "from the autobiography of Rev. Jacob Young, a Methodist minister." The "Newcastle" referred to is the present-day New Castle, the county seat of Henry County. I give the quote in full:

> The costume of the Kentuckians was a hunting shirt, buckskin pantaloons, a leathern belt around their middle, a scabbard, and a big knife fastened to their belt; some of them wore hats and some caps. Their feet were covered with moccasins, made of dressed deer skins. They did not think themselves dressed without their powder-horn and shot-pouch, or the gun and the tomahawk. They were ready, then, for all alarms. They knew but little. They could clear ground,

raise corn, and kill turkeys, deer, bears, and buffalo; and, when it became necessary, they understood the art of fighting the Indians as well as any men in the United States.

Shortly after we had taken up our residence, I was called upon to assist in opening a road from the place where Newcastle now stands, to the mouth of Kentucky river. That country, then, was an unbroken forest; there was nothing but an Indian trail passing the wilderness. I met the company early in the morning, with my axe, three days' provisions, and my knapsack. Here I found a captain, with about 100 men, all prepared to labor; about as jovial a company as I ever saw, all good-natured and civil. This was about the last of November, 1797. The day was cold and clear. The country through which the company passed was delightful; it was not a flat country, but, what the Kentuckians called, rolling ground — was quite well stored with lofty timber, and the undergrowth was very pretty. The beautiful canebrakes gave it a peculiar charm. What rendered it most interesting was the great abundance of wild turkeys, deer, bears, and other wild animals. The company worked hard all day, in quiet, and every man obeyed the captain's orders punctually.

About sundown, the captain, after a short address, told us the night was going to be very cold, and we must make very large fires. We felled the hickory trees in great abundance; made great log-heaps, mixing the dry wood with the green hickory; and, laying down a kind of sleepers under the pile, elevated the heap and caused it to burn rapidly. Every man had a water vessel in his knapsack; we searched for and found a stream of water. By this time, the fires were showing to great advantage; so we warmed our cold victuals, ate our suppers, and spent the evening in hearing the hunter's stories relative to the bloody scenes of the Indian war. We then heard some pretty fine singing, considering the circumstances.

Thus far, well; but a change began to take place. They became very rude, and raised the war-whoop. Their shrill shrieks made me tremble. They chose two captains, divided the men into two companies, and commenced fighting with the firebrands — the log heaps having burned down. The only law for their government was, that no man should throw a brand without fire on it — so that they might know how to dodge. They fought, for two or three hours, in perfect good nature; till brands became scarce, and they began to violate the law. Some were severely wounded, blood began to flow freely, and they were in a fair way of commencing a fight in earnest. At this moment, the loud voice of the captain rang out above the din, ordering every man to retire to rest. They dropped their weapons of warfare, rekindled the fires, and laid down to sleep. We finished our road according to directions, and returned home in health and peace.

* * *

The significance of this bit of history is in its utter violence. The work of clearing the road was itself violent. And from the orderly violence of that labor, these men turned for amusement to disorderly violence. They were men whose element was violence; the only alternatives they were aware of were those within the comprehension of main strength. And let us acknowledge that these were the truly influential men in the history of Kentucky, as well as in the history of most of the rest of America. In comparison to the fatherhood of such as these, the so-called "founding fathers" who established our political ideals are but distant cousins. It is not John Adams or Thomas Jefferson whom we see night after night in the magic mirror of the television set; we see these builders of the road from New Castle to the mouth of the Kentucky River. Their reckless violence has glamorized all our trivialities and evils. Their aggressions have simplified our complexities and problems. They have cut all our Gordian knots. They have appeared in all our disguises and costumes. They have worn all our uniforms. Their war whoop has sanctified our inhumanity and ratified our blunders of policy.

To testify to the persistence of their influence, it is only necessary for me to confess that I read the Reverend Young's account of them with delight; I yield a considerable admiration to the exuberance and extravagance of their fight with the firebrands; I take a certain pride in belonging to the same history and the same place that they belong to—though I know that they represent the worst that is in us, and in me, and that their presence in our history has been ruinous, and that their survival among us promises ruin.

"They knew but little," the observant Reverend says of them, and this is the most suggestive thing he says. It is surely understandable and pardonable, under the circumstances, that these men were ignorant by the standards of formal schooling. But one immediately reflects that the American Indian, who was ignorant by the same standards, nevertheless knew how to live in the country without making violence the invariable mode of his relation to it; in fact, from the ecologist's or the conservationist's point of view, he did it *no* violence. This is because he had, in place of what we would call education, a fully integrated culture, the content of which was a highly complex sense of his dependence on the earth. The same, I believe, was generally true of the peasants of certain old agricultural societies, particularly in the Orient. They belonged by an intricate awareness to the earth they lived on and by, which meant that they respected it, which meant that they practiced strict economies in the use of it.

The abilities of those Kentucky road builders of 1797 were far more primitive and rudimentary than those of the Stone Age people they had driven out. They could clear the ground, grow corn, kill game, and make war. In the minds and hands of men who "know but little"—or little else—all of these abilities are certain to be destructive, even of those values and benefits their use may be intended to serve.

On such a night as the Reverend Young describes, an Indian would have made do with a small shelter and a small fire. But these road builders, veterans of the Indian War, "felled the hickory trees in great abundance; made great log-heaps . . . and caused [them] to burn rapidly." Far from making a small shelter that could be adequately heated by a small fire, their way was to make no shelter at all, and heat instead a sizable area of the landscape. The idea was that when faced with abundance one should consume abundantly—an idea that has survived to become the basis of our present economy. It is neither natural nor civilized, and even from a "practical" point of view it is to the last degree brutalizing and stupid.

I think that the comparison of these road builders with the Indians, on the one hand, and with Old World peasants on the other, is a most suggestive one. The Indians and the peasants were people who belonged deeply and intricately to their places. Their ways of life had evolved slowly in accordance with their knowledge of their land, of its needs, of their own relation of dependence and responsibility to it. The road builders, on the contrary, were *placeless* people. That is why they "knew but little." Having left Europe far behind, they had not yet in any meaningful sense arrived in America, not yet having *devoted* themselves to any part of it in a way that would produce the intricate knowledge of it necessary to live in it without destroying it. Because they belonged to no place, it was almost inevitable that they should behave violently toward the places they came to. We *still* have not, in any meaningful way, arrived in America. And in spite of our great reservoir of facts and methods, in comparison to the deep earthly wisdom of established peoples we still know but little.

But my understanding of this curiously parabolic fragment of history will not be complete until I have considered more directly that the occasion of this particular violence was the building of a road. It is obvious that one who values the idea of community cannot speak against roads without risking all sorts of absurdity. It must be noticed, nevertheless, that the predecessor to this first road was "nothing but an Indian trail passing the wilderness"—a path. The Indians, then, who had the wisdom and the grace to live in this country for perhaps ten thousand years without destroying or damaging any of it, needed for their travels no more than a footpath; but their successors, who in a century and a half plundered the area of at least half its topsoil and virtually all of its forest, felt immediately that they had to have a road. My interest is not in the question of whether or not they *needed* the road, but in the fact that the road was then, and is now, the most characteristic form of their relation to the country.

The difference between a path and a road is not only the obvious one. A path is little more than a habit that comes with knowledge of a place. It is a sort of ritual of familiarity. As a form, it is a form of contact with a known landscape. It is not destructive. It is the perfect adaptation, through experience and familiarity, of movement to place; it obeys the natural contours; such obstacles as it meets it goes around. A road, on the other hand, even the most primitive road, embodies a resistance against the landscape. Its reason is not simply the necessity for movement, but haste. Its wish is to *avoid* contact with the landscape; it seeks so far as possible to go over the country, rather than through it; its aspiration, as we see clearly in the example of our modern freeways, is to be a bridge; its tendency is to translate place into space in order to traverse it with the least effort. It is destructive, seeking to remove or destroy all obstacles in its way. The primitive road advanced by the destruction of the forest; modern roads advance by the destruction of topography.

That first road from the site of New Castle to the mouth of the Kentucky River—lost now either by obsolescence or metamorphosis—is now being crossed and to some extent replaced by its modern descendant known as I–71, and I have no wish to disturb the question of whether or not *this* road was needed. I only want to observe that it bears no relation whatever to the country it passes through. It is a pure abstraction, built to serve the two abstractions that are the poles of our national life: commerce and expensive pleasure. It was built, not according to the lay of the land, but according to a blueprint. Such homes and farmlands and woodlands as happened to be in its way are now buried under it. A part of a hill near here that would have caused it to turn aside was simply cut down and disposed of as thoughtlessly as the pioneer road builders would have disposed of a tree. Its form is the form of speed, dissatisfaction, and anxiety. It represents the ultimate in engineering sophistication, but the crudest possible valuation of life in this world. It is as adequate a symbol of our relation to our country now as that first road was of our relation to it in 1797.

* * *

But the sense of the past also gives a deep richness and resonance to nearly everything I see here. It is partly the sense that what I now see, other men that

I have known once saw, and partly that this knowledge provides an imaginative access to what I do not know. I think of the country as a kind of palimpsest scrawled over with the comings and goings of people, the erasure of time already in process even as the marks of passage are put down. There are the ritual marks of neighborhood—roads, paths between houses. There are the domestic paths from house to barns and outbuildings and gardens, farm roads threading the pasture gates. There are the wanderings of hunters and searchers after lost stock, and the speculative or meditative or inquisitive "walking around" of farmers on wet days and Sundays. There is the spiraling geometry of the rounds of implements in fields, and the passing and returning scratches of plows across croplands. Often these have filled an interval, an opening, between the retreat of the forest from the virgin ground and the forest's return to ground that has been worn out and given up. In the woods here one often finds cairns of stones picked up out of furrows, gullies left by bad farming, forgotten roads, stone chimneys of houses long rotted away or burned.

* * *

Occasionally one stumbles into a coincidence that, like an unexpected alignment of windows, momentarily cancels out the sense of historical whereabouts, giving with an overwhelming immediacy an awareness of the reality of the past.

The possibility of this awareness is always immanent in old homesites. It may suddenly bear in upon one at the sight of old orchard trees standing in the dooryard of a house now filled with baled hay. It came to me when I looked out the attic window of a disintegrating log house and saw a far view of the cleared ridges with wooded hollows in between, and nothing in sight to reveal the date. Who was I, leaning to the window? When?

It broke upon me one afternoon when, walking in the woods on one of my family places, I came upon a gap in a fence, wired shut, but with deep-cut wagon tracks still passing through it under the weed growth and the fallen leaves. Where that thicket stands there was crop ground, maybe as late as my own time. I knew some of the men who tended it; their names and faces were instantly alive in my mind. I knew how it had been with them—how they would harness their mule teams in the early mornings in my grandfather's big barn and come to the woods-rimmed tobacco patches, the mules' feet wet with the dew. And in the solitude and silence that came upon them they would set to work, their water jugs left in the shade of bushes in the fencerows.

As a child I learned the early mornings in these places for myself, riding out in the wagons with the tobacco-cutting crews to those steep fields in the dew-wet shadow of the woods. As the day went on the shadow would draw back under the feet of the trees, and it would get hot. Little whirlwinds would cross the opening, picking up the dust and the dry "ground leaves" of the tobacco. We made a game of running with my grandfather to stand, shoulders scrunched and eyes squinched, in their middles.

Having such memories, I can acknowledge only with reluctance and sorrow that those slopes should never have been broken. Rich as they were, they were too steep. The humus stood dark and heavy over them once; the plow was its doom.

* * *

Early one February morning in thick fog and spattering rain I stood on the riverbank and listened to a towboat working its way downstream. Its engines were idling, nudging cautiously through the fog into the Cane Run bend. The end of the head barge emerged finally like a shadow, and then the second barge appeared, and then the towboat itself. They made the bend, increased power, and went thumping off out of sight into the fog again.

Because the valley was so enclosed in fog, the boat with its tow appearing and disappearing again into the muffling whiteness within two hundred yards, the moment had a curious ambiguity. It was as though I was not necessarily myself at all. I could have been my grandfather, in his time, standing there watching, as I knew he had.

2

I start down from one of the heights of the upland, the town of Port Royal at my back. It is a winter day, overcast and still, and the town is closed in itself, humming and muttering a little, like a winter beehive.

The dog runs ahead, prancing and looking back, knowing the way we are about to go. This is a walk well established with us—a route in our minds as well as on the ground. There is a sort of mystery in the establishment of these ways. Anytime one crosses a given stretch of country with some frequency, no matter how wanderingly one begins, the tendency is always toward habit. By the third or fourth trip, without realizing it, one is following a fixed path, going the way one went before. After that, one may still wander, but only by deliberation, and when there is reason to hurry, or when the mind wanders rather than the feet, one returns to the old route. Familiarity has begun. One has made a relationship with the landscape, and the form and the symbol and the enactment of the relationship is the path. These paths of mine are seldom worn on the ground. They are habits of mind, directions and turns. They are as personal as old shoes. My feet are comfortable in them.

From the height I can see far out over the country, the long open ridges of the farmland, the wooded notches of the streams, the valley of the river opening beyond, and then more ridges and hollows of the same kind.

Underlying this country, nine hundred feet below the highest ridgetops, more than four hundred feet below the surface of the river, is sea level. We seldom think of it here; we are a long way from the coast, and the sea is alien to us. And yet the attraction of sea level dwells in this country as an ideal dwells in a man's mind. All our rains go in search of it and, departing, they have carved the land in a shape that is fluent and falling. The streams branch like vines, and between the branches the land rises steeply and then rounds and gentles into the long narrowing fingers of ridgeland. Near the heads of the streams even the steepest land was not too long ago farmed and kept cleared. But now it has been given up and the woods is returning. The wild is flowing back like a tide. The arable ridgetops reach out above the gathered trees like headlands into the sea, bearing their human burdens of fences and houses and barns, crops and roads.

Looking out over the country, one gets a sense of the whole of it: the ridges and hollows, the clustered buildings of the farms, the open fields, the woods, the stock ponds set like coins into the slopes. But this is a surface sense, an exterior sense, such as you get from looking down on the roof of a house. The height is a threshold from which to step down into the wooded folds of the land, the interior, under the trees and along the branching streams.

I pass through a pasture gate on a deep-worn path that grows shallow a little way beyond, and then disappears altogether into the grass. The gate has gathered thousands of passings to and fro that have divided like the slats of a fan on either side of it. It is like a fist holding together the strands of a net.

Beyond the gate the land leans always more steeply toward the branch. I follow it down, and then bear left along the crease at the bottom of the slope. I have entered the downflow of the land. The way I am going is the way the water goes. There is something comfortable and fit-feeling in this, something free in this yielding to gravity and taking the shortest way down.

As the hollow deepens into the hill, before it has yet entered the woods, the grassy crease becomes a raw gully, and along the steepening slopes on either side I can see the old scars of erosion, places where the earth is gone clear to the rock. My people's errors have become the features of my country.

It occurs to me that it is no longer possible to imagine how this country looked in the beginning, before the white people drove their plows into it. It is not possible to know what was the shape of the land here in this hollow when it was first cleared. Too much of it is gone, loosened by the plows and washed away by the rain. I am walking the route of the departure of the virgin soil of the hill. I am not looking at the same land the firstcomers saw. The original surface of the hill is as extinct as the passenger pigeon. The pristine America that the first white man saw is a lost continent, sunk like Atlantis in the sea. The thought of what was here once and is gone forever will not leave me as long as I live. It is as though I walk knee-deep in its absence.

The slopes along the hollow steepen still more, and I go in under the trees. I pass beneath the surface. I am enclosed, and my sense, my interior sense, of the country becomes intricate. There is no longer the possibility of seeing very far. The distances are closed off by the trees and the steepening walls of the hollow. One cannot grow familiar here by sitting and looking as one can up in the open on the ridge. Here the eyes become dependent on the feet. To see the woods from the inside one must look and move and look again. It is inexhaustible in its standpoints. A lifetime will not be enough to experience it all. Not far from the beginning of the woods, and set deep in the earth in the bottom of the hollow, is a rock-walled pool not a lot bigger than a bathtub. The wall is still nearly as straight and tight as when it was built. It makes a neatly turned narrow horseshoe, the open end downstream. This is a historical ruin, dug here either to catch and hold the water of the little branch, or to collect the water of a spring whose vein broke to the surface here—it is probably no longer possible to know which. The pool is filled with earth now, and grass grows in it. And the branch bends around it, cut down to the bare rock, a torrent after heavy rain, other times bone dry. All that is certain is that when the pool was dug and walled there was deep topsoil on the hill to gather and hold the water. And this high up, at least, the bottom of the hollow, instead of the present raw notch of the streambed, wore the same mantle of soil as the slopes, and the stream was a steady seep or trickle, running most or all of the year. This tiny pool no doubt once furnished water for a considerable number of stock through the hot summers. And now it is only a lost souvenir, archaic and useless, except for the bitter intelligence there is in it. It is one of the monuments to what is lost.

Wherever one goes along the streams of this part of the country, one is apt to come upon this old stonework. There are walled springs and pools. There are the walls built in the steeper hollows where the fences cross or used to cross; the streams have drifted dirt in behind them, so that now where they are still intact they make waterfalls that have scooped out small pools at their feet. And there used to be miles of stone fences, now mostly scattered and sifted back into the ground.

Considering these, one senses a historical patience, now also extinct in the country. These walls were built by men working long days for little wages, or by slaves. It was work that could not be hurried at, a meticulous finding and fitting together, as though reconstructing a previous wall that had been broken up and scattered like puzzle pieces. The wall would advance only a few yards a day. The pace of it could not be borne by most modern men, even if the wages could be afforded. Those men had to move in closer accord with their own rhythms, and nature's, than we do. They had no machines. Their capacities were only those of flesh and blood. They talked as they worked. They joked and laughed. They sang. The work was exacting and heavy and hard and slow. No opportunity for pleasure was missed or slighted. The days and the years were long. The work was long. At the end of this job the next would begin. Therefore, be patient. Such pleasure as there is, is here, now. Take pleasure as it comes. Take work as it comes. The end may never come, or when it does it may be the wrong end.

Now the men who built the walls and the men who had them built have long gone underground to be, along with the buried ledges and the roots and the burrowing animals, a part of the nature of the place in the minds of the ones who come after them. I think of them lying still in their graves, as level as the sills and thresholds of their lives, as though resisting to the last the slant of the ground. And their old walls, too, reenter nature, collecting lichens and mosses with patience their builders never conceived.

Like the pasture gates, the streams are great collectors of comings and goings. The streams go down, and paths always go down beside the streams. For a while I walk along an old wagon road that is buried in leaves—a fragment, beginningless and endless as the middle of a sentence on some scrap of papyrus. There is a cedar whose branches reach over this road, and under the branches I find the leavings of two kills of some bird of prey. The most recent is a pile of blue jay feathers. The other has been rained on and is not identifiable. How little we know. How little of this was intended or expected by any man. The road that has become the grave of men's passages has led to the life of the woods.

> And I say to myself: Here is your road
> without beginning or end, appearing
> out of the earth and ending in it, bearing
> no load but the hawk's kill, and the leaves
> building earth on it, something more

to be borne. Tracks fill with earth
and return to absence. The road was worn
by men bearing earth along it. They have come
to endlessness. In their passing
they could not stay in, trees have risen
and stand still. It is leading to the dark,
to mornings where you are not. Here
is your road, beginningless and endless as God.

Now I have come down within the sound of the water. The winter has been rainy, and the hill is full of dark seeps and trickles, gathering finally, along these creases, into flowing streams. The sound of them is one of the elements, and defines a zone. When their voices return to the hill after their absence during summer and autumn, it is a better place to be. A thirst in the mind is quenched.

I have already passed the place where water began to flow in the little streambed I am following. It broke into the light from beneath a rock ledge, a thin glittering stream. It lies beside me as I walk, overtaking me and going by, yet not moving, a thread of light and sound. And now from below comes the steady tumble and rush of the water of Camp Branch—whose nameless camp was it named for?—and gradually as I descend the sound of the smaller stream is lost in the sound of the larger.

The two hollows join, the line of the meeting of the two spaces obscured even in winter by the trees. But the two streams meet precisely as two roads. That is, the stream*beds* do; the one ends in the other. As for the meeting of the waters, there is no looking at that. The one flow does not end in the other, but continues in it, one with it, two clarities merged without a shadow.

All waters are one. This is a reach of the sea, flung like a net over the hill, and now drawn back to the sea. And as the sea is never raised in the earthly nets of fishermen, so the hill is never caught and pulled down by the watery net of the sea. But always a little of it is. Each of the gathering strands of the net carries back some of the hill melted in it. Sometimes, as now, it carries so little that the water flows clear; sometimes it carries a lot and is brown and heavy with it. Whenever greedy or thoughtless men have lived on it, the hill has literally flowed out of their tracks into the bottom of the sea.

There appears to be a law that when creatures have reached the level of consciousness, as men have, they must become conscious of the creation; they must learn how they fit into it and what its needs are and what it requires of them, or else pay a terrible penalty: the spirit of the creation will go out of them, and they will become destructive; the very earth will depart from them and go where they cannot follow.

My mind is never empty or idle at the joinings of streams. Here is the work of the world going on. The creation is felt, alive and intent on its materials, in such places. In the angle of the meeting of the two streams stands the steep wooded point of the ridge, like the prow of an up-turned boat—finished, as it was a thousand years ago, as it will be in a thousand years. Its becoming is only incidental to its being. It will be because it is. It has no aim or end except to be. By being, it is growing and wearing into what it will be. The fork of the stream lies at the foot of the slope like hammer and chisel laid down at the foot of a finished sculpture. But the stream is no dead tool; it is alive, it is still at its work. Put your hand to it to learn the health of this part of the world. It is the wrist of the hill.

Perhaps it is to prepare to hear someday the music of the spheres that I am always turning my ears to the music of streams. There is indeed a music in streams, but it is not for the hurried. It has to be loitered by and imagined. Or imagined *toward*, for it is hardly for men at all. Nature has a patient ear. To her the slowest funeral march sounds like a jig. She is satisfied to have the notes drawn out to the lengths of days or weeks or months. Small variations are acceptable to her, modulations as leisurely as the opening of a flower.

The stream is full of stops and gates. Here it has piled up rocks in its path, and pours over them into a tiny pool it has scooped at the foot of its fall. Here it has been dammed by a mat of leaves caught behind a fallen limb. Here it must force a narrow passage, here a wider one. Tomorrow the flow may increase or slacken, and the tone will shift. In an hour or a week that rock may give way, and the composition will advance by another note. Some idea of it may be got by walking slowly along and noting the changes as one passes from one little fall or rapid to another. But this is a highly simplified and diluted version of the real thing, which is too complex and widespread ever to be actually heard by us. The ear must imagine an impossible patience in order to grasp even the unimaginableness of such music.

But the creation is musical, and this is a part of its music, as birdsong is, or the words of poets. The music of the streams is the music of the shaping of the earth, by which the rocks are pushed and shifted downward toward the level of the sea.

And now I find an empty beer can lying in the path. This is the track of the ubiquitous man Friday of all our woods. In my walks I never fail to discover some sign that he has preceded me. I find his empty shotgun shells, his empty cans and bottles, his sandwich wrappings. In wooded places along roadsides one is apt to find, as well, his overtraveled bedsprings, his outcast refrigerator, and heaps of the imperishable refuse of his modern kitchen. A year ago, almost in this same place where I have found his beer can, I found a possum that he had shot dead and left lying, in celebration of his manhood. He is the true American pioneer, perfectly at rest in his assumption that he is the first and the last whose inheritance and fate this place will ever be. Going forth, as he may think, to sow, he only broadcasts his effects.

As I go on down the path alongside Camp Branch, I walk by the edge of croplands abandoned only within my own lifetime. On my left are the south slopes where the woods is old, long undisturbed. On my right, the more fertile north slopes are covered with patches of briars and sumacs and a lot of young walnut trees. Tobacco of an extraordinary quality was once grown here, and then the soil wore thin, and these places were given up for the more accessible ridges that were not so steep, where row cropping made better sense anyway. But now, under the thicket growth, a mat of bluegrass has grown to testify to the good nature of this ground. It was fine dirt that lay here once, and I am far from being able to say that I could have resisted the temptation to plow it. My understanding of what is best for it is the tragic understanding of hindsight, the awareness that I have been taught what was here to be lost by the loss of it.

We have lived by the assumption that what was good for us would be good for the world. And this has been based on the even flimsier assumption that we could know with any certainty what was good even for us. We have fulfilled the danger of this by making our personal pride and greed the standard of our behavior toward the world—to the incalculable disadvantage of the world and every living thing in it. And now, perhaps very close to too late, our great error has become clear. It is not only our own creativity—our own capacity for life—that is stifled by our arrogant assumption; the creation itself is stifled.

We have been wrong. We must change our lives, so that it will be possible to live by the contrary assumption that what is good for the world will be good for us. And that requires that we make the effort to *know* the world and to learn what is good for it. We must learn to cooperate in its processes, and to yield to its limits. But even more important, we must learn to acknowledge that the creation is full of mystery; we will never entirely understand it. We must abandon arrogance and stand in awe. We must recover the sense of the majesty of creation, and the ability to be worshipful in its presence. For I do not doubt that it is only on the condition of humility and reverence before the world that our species will be able to remain in it.

Standing in the presence of these worn and abandoned fields, where the creation has begun its healing without the hindrance or the help of man, with the voice of the stream in the air and the woods standing in silence on all the slopes around me, I am deep in the interior not only of my place in the world, but of my own life, its sources and searches and concerns. I first came into these places following the men to work when I was a child. I knew the men who took their lives from such fields as these, and their lives to a considerable extent made my life what it is. In what came to me from them there was both wealth and poverty, and I have been a long time discovering which was which.

It was in the woods here along Camp Branch that Bill White, my grandfather's Negro hired hand, taught me to hunt squirrels. Bill lived in a little tin-roofed house on up nearer the head of the hollow. And this was, I suppose more than any other place, his hunting ground. It was the place of his freedom, where he could move without subservience, without considering who he was or who anybody else was. On late summer mornings, when it was too wet to work, I would follow him into the woods. As soon as we stepped in under the trees he would become silent and absolutely attentive to the life of the place. He was a good teacher and an exacting one. The rule seemed to be that if I wanted to stay with him, I had to make it possible for him to forget I was there. I was to make no noise. If I did he would look back and make a downward emphatic gesture with his hand, as explicit as writing: Be quiet, or go home. He would see a squirrel crouched in a fork or lying along the top of a branch, and indicate with a grin and a small jerk of his head where I should look; and then wait, while I, conscious of being watched and demanded upon, searched it out for myself. He taught me to look and to listen and to be quiet. I wonder if he knew the value of such teaching or the rarity of such a teacher.

In the years that followed I hunted often here alone. And later in these same woods I experienced my first obscure dissatisfactions with hunting. Though I could not have put it into words then, the sense had come to me that hunting as I knew it — the eagerness to kill something I did not need to eat — was an artificial relation to the place, when what I was beginning to need, just as inarticulately then, was a relation that would be necessary and meaningful. That was a time of great uneasiness and restlessness for me. It would be the fall of the year, the leaves would be turning, and ahead of me would be another year of school. There would be confusions about girls and ambitions, the wordless hurried feeling that time and events and my own nature were pushing me toward what I was going to be — and I had no notion what it was, or how to prepare.

And then there were years when I did not come here at all — when these places and their history were in my mind, and part of me, in places thousands of miles away. And now I am here again, changed from what I was, and still changing. The future is no more certain to me now than it ever was, though its risks are clearer, and so are my own desires: I am the father of two young children whose lives are hostages given to the future. Because of them and because of events in the world, life seems more fearful and difficult to me now than ever before. But it is also more inviting, and I am constantly aware of its nearness to joy. Much of the interest and excitement that I have in my life now has come from the deepening, in the years since my return here, of my relation to this place. For in spite of all that has happened to me in other places, the great change and the great possibility of change in my life has been in my sense of this place. The major difference is perhaps only that I have grown able to be wholeheartedly present here. I am able to sit and be quiet at the foot of some tree here in this woods along Camp Branch, and feel a deep peace, both in the place and in my awareness of it, that not too long ago I was not conscious of the possibility of. This peace is partly in being free of the suspicion that pursued me for most of my life, no matter where I was, that there was perhaps another place I *should* be, or would be happier or better in; it is partly in the increasingly articulate consciousness of being here, and of the significance and importance of being here.

After more than thirty years I have at last arrived at the candor necessary to stand on this part of the earth that is so full of my own history and so much damaged by it, and ask: What *is* this place? What is in it? What is its nature? How should men live in it? What must I do?

I have not found the answers, though I believe that in partial and fragmentary ways they have begun to come to me. But the questions are more important than their answers. In the final sense they *have* no answers. They are like the questions—they are perhaps the same questions—that were the discipline of Job. They are a part of the necessary enactment of humility, teaching a man what his importance is, what his responsibility is, and what his place is, both on the earth and in the order of things. And though the answers must always come obscurely and in fragments, the questions must be asked. They are fertile questions. In their implications and effects, they are moral and aesthetic and, in the best and fullest sense, practical. They promise a relationship to the world that is decent and preserving.

They are also, both in origin and effect, religious. I am uneasy with the term, for such religion as has been openly practiced in this part of the world has promoted and fed upon a destructive schism between body and soul,

Heaven and earth. It has encouraged people to believe that the world is of no importance, and that their only obligation in it is to submit to certain churchly formulas in order to get to Heaven. And so the people who might have been expected to care most selflessly for the world have had their minds turned elsewhere—to a pursuit of "salvation" that was really only another form of gluttony and self-love, the desire to perpetuate their lives beyond the life of the world. The Heaven-bent have abused the earth thoughtlessly, by inattention, and their negligence has permitted and encouraged others to abuse it deliberately. Once the creator was removed from the creation, divinity became only a remote abstraction, a social weapon in the hands of the religious institutions. This split in public values produced or was accompanied by, as it was bound to be, an equally artificial and ugly division in people's lives, so that a man, while pursuing Heaven with the sublime appetite he thought of as his soul, could turn his heart against his neighbors and his hands against the world. For these reasons, though I know that my questions *are* religious, I dislike having to *say* that they are.

But when I ask them my aim is not primarily to get to Heaven. Though Heaven is certainly more important than the earth if all they say about it is true, it is still morally incidental to it and dependent on it, and I can only imagine it and desire it in terms of what I know of the earth. And so my questions do not aspire beyond the earth. They aspire *toward* it and *into* it. Perhaps they aspire *through* it. They are religious because they are asked at the limit of what I know; they acknowledge mystery and honor its presence in the creation; they are spoken in reverence for the order and grace that I see, and that I trust beyond my power to see.

The stream has led me down to an old barn built deep in the hollow to house the tobacco once grown on those abandoned fields. Now it is surrounded by the trees that have come back on every side—a relic, a fragment of another time, strayed out of its meaning. This is the last of my historical landmarks. To here, my walk has had insistent overtones of memory and history. It has been a movement of consciousness through knowledge, eroding and shaping, adding and wearing away. I have descended like the water of the stream through what I know of myself, and now that I have there is a little more to know. But here at the barn, the old roads and the cow paths—the formal connections with civilization—come to an end.

I stoop between the strands of a barbed-wire fence, and in that movement I go out of time into timelessness. I come into a wild place. The trees grow big, their trunks rising clean, free of undergrowth. The place has a serenity and dignity that one feels immediately; the creation is whole in it and unobstructed. It is free of the strivings and dissatisfactions, the partialities and imperfections of places under the mechanical dominance of men. Here, what to a housekeeper's eye might seem disorderly is nonetheless orderly and within order; what might seem arbitrary or accidental is included in the design of the whole; what might seem evil or violent is a comfortable member of the household. Where the creation is whole nothing is extraneous. The presence of the creation here makes this a holy place, and it is as a pilgrim that I have come. It is the creation that has attracted me, its perfect interfusion of life and design. I have made myself its follower and its apprentice.

One early morning last spring, I came and found the woods floor strewn with bluebells. In the cool sunlight and the lacy shadows of the spring woods the blueness of those flowers, their elegant shape, their delicate fresh scent kept me standing and looking. I found a delight in them that I cannot describe and that I will never forget. Though I had been familiar for years with most of the spring woods flowers, I had never seen these and had not known they were here. Looking at them, I felt a strange loss and sorrow that I had never seen them before. But I was also exultant that I saw them now—that they were here.

For me, in the thought of them will always be the sense of the joyful surprise with which I found them—the sense that came suddenly to me then that the world is blessed beyond my understanding, more abundantly than I will ever know. What lives are still ahead of me here to be discovered and exulted in, tomorrow, or in twenty years? What wonder will be found here on the morning after my death? Though as a man I inherit great evils and the possibility of great loss and suffering, I know that my life is blessed and graced by the yearly flowering of the bluebells. How perfect they are! In their presence I am humble and joyful. If I were given all the learning and all the methods of my race I could not make one of them, or even imagine one. Solomon in all his glory was not arrayed like one of these. It is the privilege and the labor of the apprentice of creation to come with his imagination into the unimaginable, and with his speech into the unspeakable.

3

Sometimes I can no longer think in the house or in the garden or in the cleared fields. They bear too much resemblance to our failed human history—failed, because it has led to this human present that is such a bitterness and a trial. And so I go to the woods. As I go in under the trees, dependably, almost at once, and by nothing I do, things fall into place. I enter an order that does not exist outside, in the human spaces. I feel my life take its place among the lives—the trees, the annual plants, the animals and birds, the living of all these and the dead—that go and have gone to make the life of the earth. I am less important than I thought, the human race is less important than I thought. I rejoice in that. My mind loses its urgings, senses its nature, and is free. The forest grew here in its own time, and so I will live, suffer and rejoice, and die in my own time. There is nothing that I may decently hope for that I cannot reach by patience as well as by anxiety. The hill, which is a part of America, has killed no one in the service of the American government. Then why should I, who am a fragment of the hill? I wish to be as peaceable as my land, which does no violence, though it has been the scene of violence and has had violence done to it.

How, having a consciousness, an intelligence, a human spirit—all the vaunted equipment of my race—can I humble myself before a mere piece of the earth and speak of myself as its fragment? Because my mind transcends the hill only to be filled with it, to comprehend it a little, to know that it lives on the hill in time as well as in place, to recognize itself as the hill's fragment.

The false and truly belittling transcendence is ownership. The hill has had more owners than its owners have had years—they are grist for its mill. It has had few friends. But I wish to be its friend, for I think it serves its friends well. It tells them they are fragments of its life. In its life they transcend their years.

* * *

The most exemplary nature is that of the topsoil. It is very Christ-like in its passivity and beneficence, and in the penetrating energy that issues out of its peaceableness. It increases by experience, by the passage of seasons over it, growth rising out of it and returning to it, not by ambition or aggressiveness. It is enriched by all things that die and enter into it. It keeps the past, not as history or as memory, but as richness, new possibility. Its fertility is always building up out of death into promise. Death is the bridge or the tunnel by which its past enters its future.

* * *

To walk in the woods, mindful only of the *physical* extent of it, is to go perhaps as owner, or as knower, confident of one's own history and of one's own importance. But to go there, mindful as well of its temporal extent, of the age of it, and of all that led up to the present life of it, and of all that may follow it, is to feel oneself a flea in the pelt of a great living thing, the discrepancy between its life and one's own so great that it cannot be imagined. One has come into the presence of mystery. After all the trouble one has taken to be a modern man, one has come back under the spell of a primitive awe, wordless and humble.

* * *

In the centuries before its settlement by white men, among the most characteristic and pleasing features of the floor of this valley, and of the stream banks on its slopes, were the forests and the groves of great beech trees. With their silver bark and their light graceful foliage, turning gold in the fall, they were surely as lovely as any forests that ever grew on earth. I think so because I have seen their diminished descendants, which have returned to stand in the wasted places that we have so quickly misused and given up. But those old forests are all gone. We will never know them as they were. We have driven them beyond the reach of our minds, only a vague hint of their presence returning to haunt us, as though in dreams—a fugitive rumor of the nobility and beauty and abundance of the squandered maidenhood of our world—so that, do what we will, we will never quite be satisfied ever again to be here.

The country, as we have made it by the pretense that we can do without it as soon as we have completed its metamorphosis into cash, no longer holds even the possibility of such forests, for the topsoil that they made and stood upon, like children piling up and trampling underfoot the fallen leaves, is no longer here.

* * *

There is an ominous—perhaps a fatal—presumptuousness in living in a place by the *imposition* on it of one's ideas and wishes. And that is the way we white people have lived in America throughout our history, and it is the way our history now teaches us to live here.

Surely there could be a more indigenous life than we have. There could be a consciousness that would establish itself on a place by understanding its nature and learning what is potential in it. A man ought to study the wilderness of a place before applying to it the ways he learned in another place. Thousands of acres of hill land, here and in the rest of the country, were wasted by a system of agriculture that was fundamentally alien to it. For more than a century, here, the steepest hillsides were farmed, by my forefathers and their neighbors, as if they were flat, and as if this was not a country of heavy rains. We haven't yet, in any meaningful sense, arrived in these places that we declare we own. We undertook the privilege of the virgin abundance of this land without any awareness at all that we undertook at the same time a responsibility toward it. That responsibility has never yet impressed itself upon our character; its absence in us is signified on the land by scars.

Until we understand what the land is, we are at odds with everything we touch. And to come to that understanding it is necessary, even now, to leave the regions of our conquest—the cleared fields, the towns and cities, the highways—and re-enter the woods. For only there can a man encounter the silence and the darkness of his own absence. Only in this silence and darkness can he recover the sense of the world's longevity, of its ability to thrive without him, of his inferiority to it and his dependence on it. Perhaps then, having heard that silence and seen that darkness, he will grow humble before the place and begin to take it in—to learn *from it* what it is. As its sounds come into his hearing, and its lights and colors come into his vision, and its odors come into his nostrils, then he may come into *its* presence as he never has before, and he will arrive in his place and will want to remain. His life will grow out of the ground like the other lives of the place, and take its place among them. He will be *with* them—neither ignorant of them, nor indifferent to them, nor against them—and so at last he will grow to be native-born. That is, he must reenter the silence and the darkness, and be born again.

One winter night nearly twenty years ago I was in the woods with the coon hunters, and we were walking toward the dogs, who had moved out to the point of the bluff where the valley of Cane Run enters the valley of the river. The footing was difficult, and one of the hunters was having trouble with his lantern. The flame would "run up" and smoke the globe, so that the light it gave obscured more than it illuminated, an obstacle between his eyes and the path. At last he cursed it and flung it down into a hollow. Its little light went looping down through the trees and disappeared, and there was a distant tinkle of glass as the globe shattered. After that he saw better and went along the bluff easier than before, and lighter, too.

Not long ago, walking up there, I came across his old lantern lying rusted in the crease of the hill, half buried already in the siftings of the slope, and I let it lie. But I've kept the memory that it renewed. I have made it one of my myths of the hill. It has come to be truer to me now than it was then.

For I have turned aside from much that I knew, and have given up much that went before. What will not bring me, more certainly than before, to where I am is of no use to me. I have stepped out of the clearing into the woods. I have thrown away my lantern, and I can see the dark.

* * *

The hill, like Valéry's sycamore, is a voyager standing still. Never moving a step, it travels through years, seasons, weathers, days and nights. These are the measures of its time, and they alter it, marking their passage on it as on a man's face. The hill has never observed a Christmas or an Easter or a Fourth of July. It has nothing to do with a dial or a calendar. Time is told in it mutely and immediately, with perfect accuracy, as it is told by the heart in the body. Its time is the birth and the flourishing and the death of the many lives that are its life.

* * *

The hill is like an old woman, all her human obligations met, who sits at work day after day, in a kind of rapt leisure, at an intricate embroidery. She has time for all things. Because she does not expect ever to be finished, she is endlessly patient with details. She perfects flower and leaf, feather and song, adorning the briefest life in great beauty as though it were meant to last forever.

* * *

In the early spring I climb up through the woods to an east-facing bluff where the bloodroot bloom in scattered colonies around the foot of the rotting monument of a tree trunk. The sunlight is slanting, clear, through the leafless branches. The flowers are white and perfect, delicate as though shaped in air and water. There is a fragility about them that communicates how short a time they will last. There is some subtle bond between them and the dwindling great trunk of the dead tree. There comes on me a pressing wish to preserve them. But I know that what draws me to them would not pass over into anything I can *do*. They will be lost. In a few days none will be here.

* * *

Coming upon a mushroom growing out of a pad of green moss between the thick roots of an oak, the sun and the dew still there together, I have felt my mind irresistibly become small, to inhabit that place, leaving me standing vacant and bewildered, like a boy whose captured field mouse has just leaped out of his hand.

* * *

As I slowly fill with the knowledge of this place, and sink into it, I come to the sense that my life here is inexhaustible, that its possibilities lie rich behind and ahead of me, that when I am dead it will not be used up.

* * *

Too much that we do is done at the expense of something else, or somebody else. There is some intransigent destructiveness in us. My days, though I think

I know better, are filled with a thousand irritations, worries, regrets for what has happened and fears for what may, trivial duties, meaningless torments—as destructive of my life as if I wanted to be dead. Take today for what it is, I counsel myself. Let it be enough.

And I dare not, for fear that if I do, yesterday will infect tomorrow. We are in the habit of contention—against the world, against each other, against ourselves.

It is not from ourselves that we will learn to be better than we are.

* * *

In spite of all the talk about the law of tooth and fang and the struggle for survival, there is in the lives of the animals and birds a great peacefulness. It is not all fear and flight, pursuit and killing. That is part of it, certainly; and there is cold and hunger; there is the likelihood that death, when it comes, will be violent. But there is peace, too, and I think that the intervals of peace are frequent and prolonged. These are the times when the creature rests, communes with himself or with his kind, takes pleasure in being alive.

This morning while I wrote I was aware of a fox squirrel hunched in the sunlight on a high elm branch beyond my window. The night had been frosty, and now the warmth returned. He stayed there a long time, warming and grooming himself. Was he not at peace? Was his life not pleasant to him then?

I have seen the same peacefulness in a flock of wood ducks perched above the water in the branches of a fallen beech, preening and dozing in the sunlight of an autumn afternoon. Even while they dozed they had about them the exquisite alertness of wild things. If I had shown myself they would have been instantly in the air. But for the time there was no alarm among them, and no fear. The moment was whole in itself, satisfying to them and to me.

Or the sense of it may come with watching a flock of cedar waxwings eating wild grapes in the top of the woods on a November afternoon. Everything they do is leisurely. They pick the grapes with a curious deliberation, comb their feathers, converse in high windy whistles. Now and then one will fly out and back in a sort of dancing flight full of whimsical flutters and turns. They are like farmers loafing in their own fields on Sunday. Though they have no Sundays, their days are full of sabbaths.

* * *

One clear fine morning in early May, when the river was flooded, my friend and I came upon four rough-winged swallows circling over the water, which was still covered with frail wisps and threads of mist from the cool night. They were bathing, dipping down to the water until they touched the still surface with a little splash. They wound their flight over the water like the graceful falling loops of a fine cord. Later they perched on a dead willow, low to the water, to dry and groom themselves, the four together. We paddled the canoe almost within reach of them before they flew. They were neat, beautiful, gentle birds. Sitting there preening in the sun after their cold bath, they communicated a sense of domestic integrity, the serenity of living within order. We didn't belong within the order of the events and needs of their day, and so they didn't notice us until they had to.

* * *

But there is not only peacefulness, there is joy. And the joy, less deniable in its evidence than the peacefulness, is the confirmation of it. I sat one summer evening and watched a great blue heron make his descent from the top of the hill into the valley. He came down at a measured deliberate pace, stately as always, like a dignitary going down a stair. And then, at a point I judged to be midway over the river, without at all varying his wingbeat he did a backward turn in the air, a loop-the-loop. It could only have been a gesture of pure exuberance, of joy—a speaking of his sense of the evening, the day's fulfillment, his descent homeward. He made just the one slow turn, and then flew on out of sight in the direction of a slew farther down in the bottom. The movement was incredibly beautiful, at once exultant and stately, a benediction on the evening and on the river and on me. It seemed so perfectly to confirm the presence of a free nonhuman joy in the world—a joy I feel a great need to believe in—that I had the skeptic's impulse to doubt that I had seen it. If I had, I thought, it would be a sign of the presence of something heavenly in the earth. And then, one evening a year later, I saw it again.

* * *

Every man is followed by a shadow which is his death—dark, featureless, and mute. And for every man there is a place where his shadow is clarified and is made his reflection, where his face is mirrored in the ground. He sees his source and his destiny, and they are acceptable to him. He becomes the follower of what pursued him. What hounded his track becomes his companion.

That is the myth of my search and my return.

* * *

I have been walking in the woods, and have lain down on the ground to rest. It is the middle of October, and around me, all through the woods, the leaves are quietly sifting down. The newly fallen leaves make a dry, comfortable bed, and I lie easy, coming to rest within myself as I seem to do nowadays only when I am in the woods.

And now a leaf, spiraling down in wild flight, lands on my shirt at about the third button below the collar. At first I am bemused and mystified by the coincidence—that the leaf should have been so hung, weighted and shaped, so ready to fall, so nudged loose and slanted by the breeze, as to fall where I, by the same delicacy of circumstance, happened to be lying. The event, among all its ramifying causes and considerations, and finally its mysteries, begins to take on the magnitude of history. Portent begins to dwell in it.

And suddenly I apprehend in it the dark proposal of the ground. Under the fallen leaf my breastbone burns with imminent decay. Other leaves fall. My body begins its long shudder into humus. I feel my substance escape me, carried into the mold by beetles and worms. Days, winds, seasons pass over me as I sink under the leaves. For a time only sight is left me, a passive awareness of the sky overhead, birds crossing, the mazed interreaching of the treetops, the leaves falling—and then that, too, sinks away. It is acceptable to me, and I am at peace.

When I move to go, it is as though I rise up out of the world.

PART II

Understanding Our Cultural Crisis

An agrarian voice is rarely heard in cultural debate, even if it is acknowledged that people invariably depend on the sources of soil, water, and air for their well-being. The result is an incomplete, often skewed accounting of the roots of our cultural malaise. In the following essays Berry considers how problems of culture are intimately connected to problems in agriculture. For instance, contemporary concern over the state of the environment finds a more adequate expression and resolution if the focus of discussion moves beyond a preoccupation with wilderness areas to include the state of farming. The issues of race and gender are likewise illuminated by attention to the differing characterizations of work that follow from an agrarian ethos, an ethos that stresses the dignity of physical labor and care. Throughout this section Berry stresses that at root our cultural crises stem from a failure of character, a failure that in significant measure stems from our unwillingness to be attentive caretakers of the places we live in and from.

The Unsettling of America

One of the peculiarities of the white race's presence in America is how little intention has been applied to it. As a people, wherever we have been, we have never really intended to be. The continent is said to have been discovered by an Italian who was on his way to India. The earliest explorers were looking for gold, which was, after an early streak of luck in Mexico, always somewhere farther on. Conquests and foundings were incidental to this search—which did not, and could not, end until the continent was finally laid open in an orgy of goldseeking in the middle of the last century. Once the unknown of geography was mapped, the industrial marketplace became the new frontier, and we continued, with largely the same motives and with increasing haste and anxiety, to displace ourselves—no longer with unity of direction, like a migrant flock, but like the refugees from a broken anthill. In our own time we have invaded foreign lands and the moon with the high-toned patriotism of the conquistadors, and with the same mixture of fantasy and avarice.

That is too simply put. It is substantially true, however, as a description of the dominant tendency in American history. The temptation, once that has been said, is to ascend altogether into rhetoric and inveigh equally against all our forebears and all present holders of office. To be just, however, it is necessary to remember that there has been another tendency: the tendency to stay put, to say, "No farther. This is the place." So far, this has been the weaker tendency, less glamorous, certainly less successful. It is also the older of these tendencies, having been the dominant one among the Indians.

The Indians did, of course, experience movements of population, but in general their relation to place was based upon old usage and association, upon inherited memory, tradition, veneration. The land was their homeland. The first and greatest American revolution, which has never been superseded, was the coming of people who did *not* look upon the land as a homeland. But there were always those among the newcomers who saw that they had come to a good place and who saw its domestic possibilities. Very early, for instance, there were men who wished to establish agricultural settlements rather than quest for gold or exploit the Indian trade. Later, we know that every advance of the frontier left behind families and communities who intended to remain and prosper where they were.

But we know also that these intentions have been almost systematically overthrown. Generation after generation, those who intended to remain and prosper where they were have been dispossessed and driven out, or subverted and exploited where they were, by those who were carrying out some version of the search for El Dorado. Time after time, in place after place, these conquerors have fragmented and demolished traditional communities, the beginnings of domestic cultures. They have always said that what they destroyed was outdated, provincial, and contemptible. And with alarming frequency they have been believed and trusted by their victims, especially when their victims were other white people.

If there is any law that has been consistently operative in American history, it is that the members of any *established* people or group or community sooner or later become "redskins"—that is, they become the designated victims of an utterly ruthless, officially sanctioned and subsidized exploitation. The colonists who drove off the Indians came to be intolerably exploited by their imperial governments. And that alien imperialism was thrown off only to be succeeded by a domestic version of the same thing; the class of independent small farmers who fought the war of independence has been exploited by, and recruited into, the industrial society until by now it is almost extinct. Today, the most numerous heirs of the farmers of Lexington and Concord are the little groups scattered all over the country whose names begin with "Save": Save Our Land, Save the Valley, Save Our Mountains, Save Our Streams, Save Our Farmland. As so often before, these are *designated* victims—people without official sanction, often without official friends, who are struggling to preserve their places, their values, and their lives as they know them and prefer to live them against the agencies of their own government, which are using their own tax moneys against them.

The only escape from this destiny of victimization has been to "succeed"—that is, to "make it" into the class of exploiters, and then to remain so specialized and so "mobile" as to be unconscious of the effects of one's life or livelihood. This escape is, of course, illusory, for one man's producer is another's consumer, and even the richest and most mobile will soon find it hard to escape the noxious effluents and fumes of their various public services.

Let me emphasize that I am not talking about an evil that is merely contemporary or "modern," but one that is as old in America as the white man's presence here. It is an intention that was *organized* here almost from the start. "The New World," Bernard DeVoto wrote in *The Course of Empire*, "was a constantly expanding market. . . . Its value in gold was enormous but it had still greater value in that it expanded and integrated the industrial systems of Europe."

And he continues: "The first belt-knife given by a European to an Indian was a portent as great as the cloud that mushroomed over Hiroshima. . . . Instantly the man of 6000 B.C. was bound fast to a way of life that had developed seven and a half millennia beyond his own. He began to live better and he began to die."

The principal European trade goods were tools, cloth, weapons, ornaments, novelties, and alcohol. The sudden availability of these things produced a revolution that "affected every aspect of Indian life. The struggle for existence . . . became easier. Immemorial handicrafts grew obsolescent, then obsolete. Methods of hunting were transformed. So were methods—and the purposes—of war. As war became deadlier in purpose and armament a surplus of women developed, so that marriage customs changed and polygamy became common. The increased usefulness of women in the preparation of pelts worked to the same end. . . . Standards of wealth, prestige, and honor changed. The Indians acquired commercial values and developed business cults. They became more mobile. . . .

"In the sum it was cataclysmic. A culture was forced to change much faster than change could be adjusted to. All corruptions of culture produce breakdowns of morale, of communal integrity, and of personality, and this force was as strong as any other in the white man's subjugation of the red man."

I have quoted these sentences from DeVoto because, the obvious differences aside, he is so clearly describing a revolution that did not stop with the subjugation of the Indians, but went on to impose substantially the same catastrophe upon the small farms and the farm communities, upon the shops of small local tradesmen of all sorts, upon the workshops of independent craftsmen, and upon the households of citizens. It is a revolution that is still going on. The economy is still substantially that of the fur trade, still based on the same general kinds of commercial items: technology, weapons, ornaments, novelties, and drugs. The one great difference is that by now the revolution has deprived the mass of consumers of any independent access to the staples of life: clothing, shelter, food, even water. Air remains the only necessity that the average user can still get for himself, and the revolution has imposed a heavy tax on that by way of pollution. Commercial conquest is far more thorough and final than military defeat. The Indian became a redskin, not by loss in battle, but by accepting a dependence on traders that made *necessities* of industrial goods. This is not merely history. It is a parable.

DeVoto makes it clear that the imperial powers, having made themselves willing to impose this exploitive industrial economy upon the Indians, could not then keep it from contaminating their own best intentions: "More than four-fifths of the wealth of New France was furs, the rest was fish, and it had no agricultural wealth. One trouble was that whereas the crown's imperial policy required it to develop the country's agriculture, the crown's economy required the colony's furs, an adverse interest." And La Salle's dream of developing Louisiana (agriculturally and otherwise) was frustrated because "The interest of the court in Louisiana colonization was to secure a bridgehead for an attack on the silver mines of northern Mexico. . . ."

One cannot help but see the similarity between this foreign colonialism and the domestic colonialism that, by policy, converts productive farm, forest, and grazing lands into strip mines. Now, as then, we see the abstract values of an industrial economy preying upon the native productivity of land and people. The fur trade was only the first establishment on this continent of a mentality whose triumph is its catastrophe.

My purposes in beginning with this survey of history are (1) to show how deeply rooted in our past is the mentality of exploitation; (2) to show how fundamentally revolutionary it is; and (3) to show how crucial to our history—hence, to our own minds—is the question of how we will relate to our land.

This question, now that the corporate revolution has so determinedly invaded the farmland, returns us to our oldest crisis.

We can understand a great deal of our history—from Cortés's destruction of Tenochtitlán in 1521 to the bulldozer attack on the coalfields four-and-a-half centuries later—by thinking of ourselves as divided into conquerors and victims. In order to understand our own time and predicament and the work that is to be done, we would do well to shift the terms and say that we are divided between exploitation and nurture. The first set of terms is too simple for the purpose because, in any given situation, it proposes to divide people into two mutually exclusive groups; it becomes complicated only when we are dealing with situations in succession—as when a colonist who persecuted the Indians then resisted persecution by the crown. The terms *exploitation* and *nurture*, on the other hand, describe a division not only between persons but also within persons. We are all to some extent the products of an exploitive society, and it would be foolish and self-defeating to pretend that we do not bear its stamp.

Let me outline as briefly as I can what seem to me the characteristics of these opposite kinds of mind. I conceive a strip miner to be a model exploiter, and as a model nurturer I take the old-fashioned idea or ideal of a farmer. The exploiter is a specialist, an expert; the nurturer is not. The standard of the exploiter is efficiency; the standard of the nurturer is care. The exploiter's goal is money, profit; the nurturer's goal is health—his land's health, his own, his family's, his community's, his country's. Whereas the exploiter asks of a piece of land only how much and how quickly it can be made to produce, the nurturer asks a question that is much more complex and difficult: What is its carrying capacity? (That is: How much can be taken from it without diminishing it? What can it produce *dependably* for an indefinite time?) The exploiter wishes to earn as much as possible by as little work as possible; the nurturer expects, certainly, to have a decent living from his work, but his characteristic wish is to work *as well* as possible. The competence of the exploiter is in organization; that of the nurturer is in order—a human order, that is, that accommodates itself both to other order and to mystery. The exploiter typically serves an institution or organization; the nurturer serves land, household, community, place. The exploiter thinks in terms of numbers, quantities, "hard facts"; the nurturer in terms of character, condition, quality, kind.

It seems likely that all the "movements" of recent years have been representing various claims that nurture has to make against exploitation. The women's movement, for example, when its energies are most accurately placed, is arguing the cause of nurture; other times it is arguing the right of women to be exploiters—which men have no *right* to be. The exploiter is clearly the prototype of the "masculine" man—the wheeler-dealer whose "practical" goals require the sacrifice of flesh, feeling, and principle. The nurturer, on the other hand, has always passed with ease across the boundaries of the so-called sexual roles. Of necessity and without apology, the preserver of seed, the planter, becomes midwife and nurse. Breeder is always metamorphosing into brooder and back again. Over and over again, spring after spring, the questing mind, idealist and visionary, must pass through the planting to become nurturer of the real. The farmer, sometimes known as husbandman, is by definition half mother; the only question is how good a mother he or she is. And the land itself is not mother or father only, but both. Depending on crop and season, it is at one time receiver of seed, bearer and nurturer of young; at another, raiser of seed-stalk, bearer and shedder of seed. And in response to these changes, the farmer crosses back and forth from one zone of spousehood to another, first as planter and then as gatherer. Farmer and land are thus involved in a sort of dance in which the partners are always at opposite sexual poles, and the lead keeps changing: the farmer, as seed-bearer, causes growth; the land, as seed-bearer, causes the harvest.

The exploitive always involves the abuse or the perversion of nurture and ultimately its destruction. Thus, we saw how far the exploitive revolution had penetrated the official character when our recent secretary of agriculture remarked that "food is a weapon." This was given a fearful symmetry indeed when, in discussing the possible use of nuclear weapons, a secretary of defense spoke of "palatable" levels of devastation. Consider the associations that have since ancient times clustered around the idea of food—associations of mutual care, generosity, neighborliness, festivity, communal joy, religious ceremony—and you will see that these two secretaries represent a cultural catastrophe. The concerns of farming and those of war, once thought to be diametrically opposed, have become identical. Here we have an example of men who have been made vicious, not presumably by nature or circumstance, but by their *values*.

Food is *not* a weapon. To use it as such—to foster a mentality willing to use it as such—is to prepare, in the human character and community, the destruction of the sources of food. The first casualties of the exploitive revolution are character and community. When those fundamental integrities are devalued and broken, then perhaps it is inevitable that food will be looked upon as a weapon, just as it is inevitable that the earth will be looked upon as fuel and people as numbers or machines. But character and community—that is, culture in the broadest, richest sense—constitute, just as much as nature, the source of food. Neither nature nor people alone can produce human sustenance, but only the two together, culturally wedded. The poet Edwin Muir said it unforgettably:

> Men are made of what is made,
> The meat, the drink, the life, the corn,
> Laid up by them, in them reborn.
> And self-begotten cycles close
> About our way; indigenous art
> And simple spells make unafraid
> The haunted labyrinths of the heart
> And with our wild succession braid
> The resurrection of the rose.

To think of food as a weapon, or of a weapon as food, may give an illusory security and wealth to a few, but it strikes directly at the life of all.

The concept of food-as-weapon is not surprisingly the doctrine of a Department of Agriculture that is being used as an instrument of foreign political and economic speculation. This militarizing of food is the greatest threat so far raised against the farmland and the farm communities of this country. If present attitudes continue, we may expect government policies that will encourage the destruction, by overuse, of farmland. This, of course, has already begun. To answer the official call for more production—evidently to be used to bait or bribe foreign countries—farmers are plowing their waterways and permanent pastures; lands that ought to remain in grass are being planted in row crops. Contour plowing, crop rotation, and other conservation measures seem to have gone out of favor or fashion in official circles and are practiced less and less on the farm. This exclusive emphasis on production will accelerate the mechanization and chemicalization of farming, increase the price of land, increase overhead and operating costs, and thereby further diminish the farm population. Thus the tendency, if not the intention, of Mr. Butz's confusion of farming and war is to complete the deliverance of American agriculture into the hands of corporations.

The cost of this corporate totalitarianism in energy, land, and social disruption will be enormous. It will lead to the exhaustion of farmland and farm culture. Husbandry will become an extractive industry; because maintenance will entirely give way to production, the fertility of the soil will become a limited, unrenewable resource like coal or oil.

This may not happen. It *need* not happen. But it is necessary to recognize that it *can* happen. That it can happen is made evident not only by the words of such men as Mr. Butz, but more clearly by the large-scale industrial destruction of farmland already in progress. If it does happen, we are familiar enough with the nature of American salesmanship to know that it will be done in the name of the starving millions, in the name of liberty, justice, democracy, and brotherhood, and to free the world from communism. We must, I think, be prepared to see, and to stand by, the truth: that the land should not be destroyed for *any* reason, not even for any apparently good reason. We must be prepared to say that enough food, year after year, is possible only for a limited number of people, and that this possibility can be preserved only by the steadfast, knowledgeable *care* of those people. Such "crash programs" as apparently have been contemplated by the Department of Agriculture in recent years will, in the long run, cause more starvation than they can remedy.

Meanwhile, the dust clouds rise again over Texas and Oklahoma. "Snirt" is falling in Kansas. Snowdrifts in Iowa and the Dakotas are black with blown soil. The fields lose their humus and porosity, become less retentive of water, depend more on pesticides, herbicides, chemical fertilizers. Bigger tractors become necessary because the compacted soils are harder to work—and their greater weight further compacts the soil. More and bigger machines, more chemical and methodological shortcuts are needed because of the shortage of manpower on the farm—and the problems of overcrowding and unemployment increase in the cities. It is estimated that it now costs (by erosion) two bushels of Iowa topsoil to grow one bushel of corn. It is variously estimated that from five to twelve calories of fossil fuel energy are required to produce one calorie of hybrid corn energy. An official of the National Farmers Union says that "a farmer who earns $10,000 to $12,000 a year typically leaves an estate valued at about $320,000"—which means that when that farm is financed again, either by a purchaser or by an heir (to pay the inheritance taxes), it simply cannot support its new owner and pay for itself. And the *Progressive Farmer* predicts the disappearance of 200,000 to 400,000 farms each year during the next twenty years if the present trend continues.

* * *

The first principle of the exploitive mind is to divide and conquer. And surely there has never been a people more ominously and painfully divided than we are—both against each other and within ourselves. Once the revolution of exploitation is under way, statesmanship and craftsmanship are gradually replaced by salesmanship.* Its stock in trade in politics is to sell despotism and avarice as freedom and democracy. In business it sells sham and frustration as luxury and satisfaction. The "constantly expanding market" first opened in the New World by the fur traders is still expanding—no longer so much by expansions of territory or population, but by the calculated outdating, outmoding, and degradation of goods and by the hysterical self-dissatisfaction of consumers that is indigenous to an exploitive economy.

This gluttonous enterprise of ugliness, waste, and fraud thrives in the disastrous breach it has helped to make between our bodies and our souls. As a people, we have lost sight of the profound communion—even the union—of the inner with the outer life. Confucius said: "If a man have not order within him/He can not spread order about him. . . ." Surrounded as we are by evidence of the disorders of our souls and our world, we feel the strong truth in those words as well as the possibility of healing that is in them. We see the likelihood that our surroundings, from our clothes to our countryside, are the products of our inward life—our spirit, our vision—as much as they are products of nature and work. If this is true, then we cannot live as we do and be as we would like to be. There is nothing more absurd, to give an example that is only apparently trivial, than the millions who wish to live in luxury and idleness and yet be slender and good-looking. We have millions, too, whose livelihoods, amusements, and comforts are all destructive, who nevertheless wish to live in a healthy environment; they want to run their recreational engines in clean, fresh air. There is now, in fact, no "benefit" that is not associated with disaster. That is because power can be disposed morally or harmlessly only by thoroughly unified characters and communities.

What caused these divisions? There are no doubt many causes, complex both in themselves and in their interaction. But pertinent to all of them, I think, is our attitude toward work. The growth of the exploiters' revolution on this continent has been accompanied by the growth of the idea that work is beneath human dignity, particularly any form of hand work. We have made it our overriding ambition to escape work, and as a consequence have debased work until it is only fit to escape from. We have debased the products of work and have been, in turn, debased by them. Out of this contempt for work arose

*The craft of persuading people to buy what they do not need, and do not want, for more than it is worth.

the idea of a nigger: at first some person, and later some thing, to be used to relieve us of the burden of work. If we began by making niggers of people, we have ended by making a nigger of the world. We have taken the irreplaceable energies and materials of the world and turned them into jimcrack "labor-saving devices." We have made of the rivers and oceans and winds niggers to carry away our refuse, which we think we are too good to dispose of decently ourselves. And in doing this to the world that is our common heritage and bond, we have returned to making niggers of people: we have become each other's niggers.

But is work something that we have a right to escape? And can we escape it with impunity? We are probably the first entire people ever to think so. All the ancient wisdom that has come down to us counsels otherwise. It tells us that work is necessary to us, as much a part of our condition as mortality; that good work is our salvation and our joy; that shoddy or dishonest or self-serving work is our curse and our doom. We have tried to escape the sweat and sorrow promised in Genesis—only to find that, in order to do so, we must forswear love and excellence, health and joy.

Thus we can see growing out of our history a condition that is physically dangerous, morally repugnant, ugly. Contrary to the blandishments of the salesmen, it is not particularly comfortable or happy. It is not even affluent in any meaningful sense, because its abundance is dependent on sources that are being rapidly exhausted by its methods. To see these things is to come up against the question: Then what *is* desirable?

One possibility is just to tag along with the fantasists in government and industry who would have us believe that we can pursue our ideals of affluence, comfort, mobility, and leisure indefinitely. This curious faith is predicated on the notion that we will soon develop unlimited new sources of energy: domestic oil fields, shale oil, gasified coal, nuclear power, solar energy, and so on. This is fantastical because the basic cause of the energy crisis is not scarcity; it is moral ignorance and weakness of character. We don't know *how* to use energy, or what to use it *for*. And we cannot restrain ourselves. Our time is characterized as much by the abuse and waste of human energy as it is by the abuse and waste of fossil fuel energy. Nuclear power, if we are to believe its advocates, is presumably going to be well used by the same mentality that has egregiously devalued and misapplied man- and womanpower. If we had an unlimited supply of solar or wind power, we would use that destructively, too, for the same reasons.

Perhaps all of those sources of energy are going to be developed. Perhaps all of them can sooner or later be developed without threatening our survival. But not all of them together can guarantee our survival, and they cannot define what is desirable. We will not find those answers in Washington, D.C., or in the laboratories of oil companies. In order to find them, we will have to look closer to ourselves.

I believe that the answers are to be found in our history: in its until now subordinate tendency of settlement, of domestic permanence. This was the ambition of thousands of immigrants; it is formulated eloquently in some of the letters of Thomas Jefferson; it was the dream of the freed slaves; it was written into law in the Homestead Act of 1862. There are few of us whose families have not at some time been moved to see its vision and to attempt to enact its possibility. I am talking about the idea that as many as possible should share in the ownership of the land and thus be bound to it by economic interest, by the investment of love and work, by family loyalty, by memory and tradition. How much land this should be is a question, and the answer will vary with geography. The Homestead Act said 160 acres. The freedmen of the 1860s hoped for forty. We know that, particularly in other countries, families have lived decently on far fewer acres than that.

The old idea is still full of promise. It is potent with healing and with health. It has the power to turn each person away from the big-time promising and planning of the government, to confront in himself, in the immediacy of his own circumstances and whereabouts, the question of what methods and ways are best. It proposes an economy of necessities rather than an economy based upon anxiety, fantasy, luxury, and idle wishing. It proposes the independent, free-standing citizenry that Jefferson thought to be the surest safeguard of democratic liberty. And perhaps most important of all, it proposes an agriculture based upon intensive work, local energies, care, and long-living communities—that is, to state the matter from a consumer's point of view: a dependable, long-term food supply.

This is a possibility that is obviously imperiled—by antipathy in high places, by adverse public fashions and attitudes, by the deterioration of our present farm communities and traditions, by the flawed education and the inexperience of our young people. Yet it alone can promise us the continuity of attention and devotion without which the human life of the earth is impossible.

Sixty years ago, in another time of crisis, Thomas Hardy wrote these stanzas:

Only a man harrowing clods
In a slow silent walk
With an old horse that stumbles and nods
Half asleep as they stalk.

Only thin smoke without flame
From the heaps of couch-grass;
Yet this will go onward the same
Though Dynasties pass.

Today most of our people are so conditioned that they do not wish to harrow clods either with an old horse or with a new tractor. Yet Hardy's vision has come to be more urgently true than ever. The great difference these sixty years have made is that, though we feel that this work *must* go onward, we are not so certain that it will. But the care of the earth is our most ancient and most worthy and, after all, our most pleasing responsibility. To cherish what remains of it, and to foster its renewal, is our only legitimate hope.

Racism and the Economy

When I wrote *The Hidden Wound* in 1968, I did not see how the freedom and prosperity of the people could be separated as issues from the issue of the health of the land, and I still do not. I wrote the book because it seemed to me that the psychic wound of racism had resulted inevitably in wounds in the land, the country itself. I believed then, and I believe more strongly now, that the root of our racial problem in America is not racism. The root is in our inordinate desire to be superior—not to some inferior or subject people, though this desire leads to the subjection of people—but to our condition. We wish to rise above the sweat and bother of taking care of anything—of ourselves, of each other, or of our country. We did not enslave African blacks because they were black, but because their labor promised to free us of the obligations of stewardship, and because they were unable to prevent us from enslaving them. They were economically valuable and militarily weak.

It seems likely, then, that what we now call racism came about as a justification of slavery after the fact, not as its cause. We decided that blacks were inferior in order to persuade ourselves that it was all right to enslave them. That this is true is suggested by our present treatment of other social groups to whom we assign the laborious jobs of caretaking. For it is not only the racial minorities who receive our indifference or contempt, but economic or geographic minorities as well. Anyone who has been called "redneck" or "hillbilly" or "hick" or sometimes even "country person" or "farmer" shares with racial minorities the experience of a stigmatizing social prejudice. And such terms as "redneck" and "hillbilly" and "hick" have remained acceptable in public use long after the repudiation of such racial epithets as "nigger" and "greaser." "Rednecks" and "hillbillies" and "hicks" are scorned because they do what used to be known as "nigger work"—work that is fundamental and inescapable. And it should not be necessary to point out the connection between the oppression of women and the general contempt for household work. It is well established among us that you may hold up your head in polite society with a public lie in your mouth or other people's money in your pocket or innocent blood on your hands, but not with dishwater on your hands or mud on your shoes.

What we did not understand at the time of slavery, and understand poorly still, is that this presumption of the inferiority of economic groups is a contagion that we cannot control, for the presumed inferiority of workers inevitably infects the quality of their work, which inevitably infects the quality of the workplace, which is to say the quality of the country itself. When a nation determines that the work of providing and caretaking is "nigger work" or work for "hillbillies" or "rednecks"—that is, fundamental, necessary, inescapable, *and inferior*—then it has implanted in its own soul the infection of its ruin.

The opprobrium implied in the term "nigger work" was, of course, not a problem confined below the Ohio River. The overriding aim of Yankee ingenuity, to this day, has been "freedom from drudgery." The great motive and the great "selling point" of industrialism has been "less work." Our national goal, indeed, has been less work, and we have succeeded. Most people who work are now working less or with less effort (and skill) than they once did, and increasing numbers are not working at all.

That this "less work" has inescapably implied poorer work, poorer products, and unhealthy side effects is a fact not yet on the agenda of presidential debate. The great persons of politics are no better equipped than the average citizen to compute the costs of "less work," though they, like the average citizen, must spend their leisure time in breathing poisonous air.

The problem of race, nevertheless, is generally treated as if it could be solved merely by recruiting more blacks and other racial minorities into colleges and then into high-paying jobs. This is to assume, simply, that we can solve the problems of racial minorities by elevating them to full partnership in the problems of the racial majority. We assume that when a young black person acquires a degree, puts on a suit, and achieves a sit-down job with a corporation, the problem is to that extent solved.

The larger, graver, more dangerous problem, however, is that we have thought of no better way of solving the race problem. The "success" of the black corporate executive, in fact, only reveals the shallowness, the jeopardy, and the falseness of the "success" of the white corporate executive. This "success" is a private and highly questionable settlement that does not solve, indeed does not refer to, the issues associated with American racism. It only assumes that American blacks will be made better or more useful or more secure by becoming as greedy, selfish, wasteful, and thoughtless as affluent American whites. The aims and standards of the oppressors become the aims and standards of the oppressed, and so our ills and evils survive our successive "liberations."

The problems associated with racism, as I have already suggested, are deeply involved in our national character, and they will not be solved by a racially equitable distribution of college degrees and professional salaries. There are several of these problems, and all of them are difficult. There would be more hope of solving them if they could be understood in such a way as to show the unlikelihood that the solutions can be simple—the unlikelihood, that is, that remedies can be thought up by the people at the top and bestowed or imposed upon the people lower down. To that end, I would like to try my hand at a description of four of these problems as they appear to me.

* * *

The first problem is the displacement of the racial problem itself from the country to the cities. The story of the black race in America began and went on for more than three hundred years as a story that was mainly rural and agricultural. With the industrialization of agriculture and the increasing availability of factory jobs during and after World War II, the story of American blacks rapidly became an urban one. If all the black emigrants from the countryside had found secure jobs and agreeable dwellings in the cities, their story would, of course, have been a different and a better one. Too often, what they found were poor jobs or no jobs, and deplorable living conditions. The move from country to city, moreover, deprived them of their competence in doing for themselves. It is no exaggeration to say that, in the country, most blacks were skilled in the arts of make-do and subsistence. If most of them were poor, they were *competently* poor; they could do for themselves and for each other. They knew how to grow and harvest and prepare food. They knew how to gather wild fruits, nuts, and herbs. They knew how to hunt and fish. They knew how to use the things that their white "superiors" threw away or disregarded or overlooked. Some of them were becoming capable small landowners. In the cities, all of this know-how was suddenly of no value, and they became abjectly and dependently poor as they never had been before. In the country, despite the limits placed upon them by segregation and poverty, they possessed a certain freedom in their ability to *do* things, but once they were in the city freedom was inescapably associated with the ability to *buy* things.

This loss of the efficacy of competence would be regrettable, indeed dangerous, whatever people it might have happened to (and, of course, it has happened to white people also), but in the story of American blacks it involves a particularly poignant loss of an opportunity for justice. Since blacks had been farm workers throughout their history in America, first as slaves and then as poor sharecroppers or day laborers, the correct and appropriate justice to them would have been to help and encourage them, so far as their individual abilities and desires allowed, to become owners of small farms. This would have been the healing of the wound of slavery that the freed slaves themselves envisioned in their plea for "forty acres and a mule." Instead, they were regarded as "excess population" in the country as soon as they were replaceable by machines, and they moved into the urban slums where, still, they are regarded as "excess population." We have probably dealt with this "excess" as well as we can by moving it around, and that is only to say that we have not dealt with it at all. To deal with it, we have to understand why these people became an "excess" in the first place, and what they and we have lost in the loss of their usefulness. And in order to understand that, we have to understand the abuses, excesses, penalties, and costs of our present ways of farming. We have to understand the substitution of industrial methods and devices for human skill and human labor, and the complex costs of that substitution.

* * *

One of the costs is dispossession, which is the second problem. In 1920 black American farmers owned 916,000 farms, totaling fifteen million acres. By 1988, according to the New York Times News Service, the number of farms owned by blacks had fallen to 30,000, totaling about three million acres. Congressman Mike Espy of Mississippi says that there were 164,000 black farmers in his state in 1910, and in 1980 there were fewer than 9,000. These figures greatly embarrass some of our conventional assumptions about racism and civil rights. For one thing, the steepest decline in land ownership by blacks, which occurred from 1950 to 1970, coincides roughly with the period of their greatest gains in civil rights. For another, even though the decline of land ownership among blacks has been greater than that among whites, the decline has been precipitous and catastrophic for both races, and for both races the causes have been mostly the same. In the decline of black farming, racial prejudice has, of course, played a significant part. But, beyond that, black farmers have failed or quit because of the same economic, political, and social adversities that have affected white farmers. Black farmers, one gathers, have lost out to a considerable extent not because they were black farmers but because they were small farmers. And it is reasonable to suppose that the black small farmers have not received government help because any measures that would have helped them would also have helped the more numerous white small farmers—thus requiring official sanction for a democratic distribution of usable property.

But at least since the time of the notorious Committee for Economic Development in the early 1940s, our government's policy for rural citizens, hence ultimately for all citizens, has been dispossession: the removal of the vast majority of all races from the independent use or ownership of land or other usable property. The decline of black farmers, then, though it may be attributable to some extent to racial prejudice, is mainly attributable to their being caught up along with white farmers in the much more comprehensive and powerful prejudice against the small landowner.

As a result of this prejudice, nearly our whole population is now dispossessed, and our most populous economic classes are the affluent dispossessed and the impoverished dispossessed. How much more secure the affluent dispossessed will prove to be, in the long run, than the impoverished dispossessed is a question that we are leaving to history to decide. At present, from the point of view of the affluent, dispossession gives the advantage of freedom from the work and worry of taking care of property; the dependency that dispossession also involves is not noticed. From the point of view of the impoverished, the dependency is noticeable immediately; for them, dispossession means, simply, the loss of the ability to help themselves.

* * *

Thus, moving a problem does not correct it, but makes it much more difficult to correct. With movement, the problem changes. The black slaves and former slaves of a farming economy, whatever their political status, had an economic usefulness and value as workers, and they had in themselves a sure, if limited, competence to help themselves. Their abilities gave them a connection to the productive capacities of the country itself that was not absolutely dependent upon their employers, and those blacks who came to own land were as independent of employers as anyone else who owned land. As competent country people, they were not completely at the mercy of their political or economic condition.

Their descendents, living in the inner-city slums of the 1980s, are no longer legally excluded from the institutions of citizenship, and so their political status may be said to have improved; but their economic status has become more dependent, consumptive, and degraded than it was before. They have no direct connection to the productive capacities of the country itself, hence no ability to perform within the legitimate economy on their own behalf, and many of them are not employed at all. According to *The Washington Spectator* of April 1, 1988, "The unemployment rate among young blacks in Harlem is more than 40 percent." Many in that 40 percent undoubtedly belong to that new American class, the "permanently unemployable." As citizens they have the right to vote and such, but the unemployed and "permanently unemployable" live outside the country's economy, or are merely, so to speak, its patients.

The transition from slave to citizen is good. But the transition from useful and therefore valuable slave to useless and therefore costly economic dependent is a bewilderment. There must be some good in it, but how much of that good is net? There is certainly danger in it, for the unemployed and the "permanently unemployable" and for everyone else. It may prove to be as dangerous for all concerned as slavery was.

And how much sense, after all, does this unemployment make? People are unemployed, for one reason, because they are replaceable by machines that work more cheaply than people (in a time of cheap energy), thus enlarging the profits of the affluent class. And yet no one who has looked can argue that the country is well cared for, either its rural places or its cities. A lot of work needs to be done, work that would help the country and everyone in it, if only it were possible to pay people to do it. Forty percent of the young blacks in Harlem are unemployed, and yet the crops cannot be harvested in many areas without illegal immigrants from Mexico—two facts that cannot be put together in such a way as to make sense.

We come, then, to this question: How *racially* significant is it that we now have many blacks who have college degrees, wear suits, are members of professions or officers of corporations, if 40 percent of the young black people of Harlem are unemployed? It certainly does not mean that the black *race* is succeeding in this country. It can only mean that the black race is now as divided within and against itself as is the white race. If this is a racial achievement, its significance is not clear to me.

Or what is the racial significance of the affluence of some blacks, when many other blacks are working at the most menial jobs? It would be clearer to say that this phenomenon has an economic significance, and that its economic significance is about the same as that of slavery: as long as there are some people who wish to believe and are economically empowered to believe that they are too good to do their own work and clean up after themselves, then somebody else is going to have to do the work and the cleaning up. In an exploitive economy, there is what we might call a "nigger factor" that will remain more or less constant. If some people grow rich by making things to throw away, then many other people will have to empty the garbage cans and make the trip to the dump.

* * *

A third problem, therefore, has to do with the extreme doubtfulness of economic solutions made within the terms of an economy based on the exploitation of power, the exhaustion of natural sources, the misuse of people, and the waste of products. What, in the first place, have we gained as a nation by paying high wages to workers for the carting away of our so-called wastes and by paying much higher wages to the corporate executives who are responsible for the existence of those wastes, both kinds of employment being utterly degrading of both humanity and the world? And in the second place, how can blacks be elevated in security or dignity by an "equal" participation in such work and such earning?

In light of the issue of the spiritual health of human beings, the issue of wages may be more or less arbitrary or irrelevant. What would be a just wage for a life of carrying off other people's cans and bottles? A million dollars a year would not be enough, because such a job can be performed only by the forfeiture of the effective life of the spirit in this world. Such work is not, in the usual sense, an accomplishment. It is not productive work. The only conceivable standard for it is quantitative; it can be done thoroughly or not; one can haul off either all the cans and bottles or only some of them. It is work that by its nature cannot be good work; though it can be done carefully, it cannot be well done. There is no art in it, no science, and no skill. Its only virtue is in its necessity. But it is necessary only for a bad reason: the manufacture of "disposable" (that is, virtually worthless) products. The people for whom this work is done will be made unhappy or unhealthy if it is not done. So long as it is done, they will scarcely think of it. It is work, then, that is entirely negative in its value. Its most desirable result is to leave no visible trace.

I said and meant that a million dollars a year would not be enough for such work, and I can easily imagine the outrage with which some readers will respond to that judgment. Why, a million dollars a year is an executive's wages! And an executive, after all, has a college degree, lives in an expensive house, and drives an expensive car. He or she wears a suit and sweats only when jogging (or worrying). Since executives do no manual or menial or domestic work, an executive's body becomes only passively dirty—by its own secretions and by the filth that falls upon it out of the air; it is never soiled by any dust that *it* has raised. Obviously, then, an executive is much superior to a garbage collector. But is this true? Let us see.

An executive, of course, is understood to be a large operator, a person of the big time. A small businessman or businesswoman, the owner of a small independent store or shop, will not think him- or herself, or be thought, an executive. Such people *do;* they do not "execute." The executive deals in large quantities of products that, typically, are purchased as cheaply as possible, and sold as dearly as possible. Typically, the products are never touched by the executive, and they come from and go to people and places the executive does not know, or care about, or give any respect or allegiance to. Many of those products are not necessary. Many of them are overpriced. Many of them cause environmental or social or cultural damage. Many of them are destroyed quickly in use but remain indestructible as garbage or pollutants. Many of them are shoddily made. Many executives grow rich by the manufacture and sale of products that, being rich, they disdain to use. All of them grow rich by work that they do not do, and would disdain to do.

The work of the executive is thus as unproductive and as spiritually desolate as that of the garbage collector. Indeed, depending upon the toxicity and persistence of the products and by-products, it may be more so. Certainly, by any standard, to haul garbage away is more virtuous than to manufacture it.

What all of us—black and white—must understand is that the existence of industrial executives, as we now have them, implies inevitably the "nigger work" of garbage collectors and other menial laborers, as we now have them. The career of the black executive implies just as much "nigger work" as the career of the white executive. And the degradation of this trade in careers and souls is not limited to people. For the garbage cannot be hauled out of the world; it must be *put* somewhere. The inevitably misnamed "sanitary landfills" were once places of dignity, woodlands or marshes or farms, the homes of creatures, and now they have been made niggers also.

Nor is that the end of it. The existence of the typical corporate executive, black or white, implies inevitably not just the "nigger work" of cleaning up after other people, and the niggerfication of the people, black or white, who must deal with the messes, and of the places where the messes are hidden away; it implies also the unemployed young people and the "permanently unemployable" of the urban ghettos, who do not have the dignity even of "nigger work." The willingness to profit from a destructive economy at the top results in economic nonentity at the bottom. Economic nonentity, as we know, is a condition that people grow extremely tired of, and when tired of it become extremely dangerous.

* * *

But even from the most selfish point of view, the success of the typical executive, black or white, is not very successful, and his or her security is not very secure. For one thing, this success and security can be achieved only by investing one's life in an economy that is destroying its natural sources and therefore itself as surely as water runs downhill. For another, this personal success and security, which are usually involved in the success and security of a corporation, in no way involve the success and security of society. The terms of this success and security are individualistic and competitive. The executive, that is, takes what he or she can get by the use of whatever power is available, just as the garbage collector does. The process of gaining this success and security thus isolates the individual both from nature and from other humans—which, of course, is a description of failure and insecurity.

There are two ways by which individual success and security can be made (within mortal limits) successful and secure: they must rest on a sound understanding and practice of economic justice; and they must involve and be involved in the success and security of the community. The competitive principle excludes both of these ways.

We might as well admit that we do not have a working concept of economic justice. We are resigned to the poor principle that people earn what they earn by power, not by the quality or usefulness of their work. Insurance executives, doctors, lawyers, mechanics, factory workers, and garbage collectors all earn in proportion to their power. People such as the small farmers, who have no power, must resign themselves to earning what they can get. This is what we mean by our understanding that the market is the ultimate arbiter of economic values. Workers will not be paid according to the quality of their work or their products, but according to their power. The market is thus detached utterly from the issue of quality and made utterly subject to manipulation by the most powerful in their own interest.

The first principle of economic justice, however, is that good work will be well paid. It follows that the first necessity of economic justice is good work—something that we, as a nation, are less and less capable of doing. The market as a mere brokerage of economic power—apart from the principled high standards of the seller and the discriminating judgment of the buyer—will inevitably have a degenerative influence on both the quality of people and the quality of products.

But surely we must go even further and say that a market will be degenerative if it is not under the rule of the virtues. The most obvious lesson of slavery, one that we have never learned, is about the limits of a mere market. A mere market cannot adequately recognize or protect the full value of a creature, as seller or as buyer or as merchandise. We now call a market "free" to the extent that buyers and sellers are able to ignore this limitation. But it was a limit not ignorable by slaves or by the enemies of slavery. To them it was plain that the market was inevitably reductive: it treated people as bodies, not as souls.

We cannot now legally own the body of another person. And yet our market for labor, as for things, is more crudely reductive now than it was then. Slaves at least were priced according to their qualities and abilities, whereas now workers, in both the trades and the professions, are more and more likely to receive "the going rate," regardless of their competence. And now their work, whether physical or mental, is likely to be more degrading spiritually than the work of skilled slaves. This, paradoxically, is the result of the market's general depreciation of all physical and material things. For we have kept institutionalized in our economic system a dualism much like the dualism that justified and enforced slavery. Despite the physical force that it requires, a slave economy is under the domination of mind. Any healthy body, as we know, is able to do the work necessary for its own maintenance. It was the mind of the master, not his body, that required the service of a slave. The slave subserved an economic idea.

Work, in our own day, is on the same terms increasingly slavish, because our economy is more mind-dominated now than it was in the time of slavery, and is increasingly so. As I have already suggested, it is not necessary for executives ever to touch either the raw material or the manufactured product by which they earn money. The work of executives is entirely mental; their physical lives are artificial, given to purely consumptive activities like golf and jogging.

And the work of the tradesman or laborer or factory worker, though it deals with material things, tends to be as mind-dominated and abstract as that of the executive. The industrial laborer subserves an economic idea instituted in machines and in mechanized procedures. This is as far as possible from the work of the traditional craftsman or artist, whose making has never resembled what we now call "manufacture," but is a cooperation and conversation of mind and body and idea and material. The true craftsman does not waste materials because his or her art involves respect for materials. And the craftsman's products are not wasted because by their quality and durability they earn respect.

A dualistic society dominated by mind involves a number of dangers, of which the degradation and destruction of the material world is only the most obvious.

It is not so obvious, or so expectable, that in a mind-dominated society, fewer and fewer people will possess independently the power or ability to make up their own minds. This is because dominance of mind always implies, politically and economically, dominance by somebody else's mind—or, worse, by the "mind" of a government or corporation.

In a society in which nearly everybody is dominated by somebody else's mind or by a disembodied mind, it becomes increasingly difficult to learn the truth about the activities of governments and corporations, about the quality or value of products, or about the health of one's own place and economy.

In such a society, also, our private economies will depend less and less upon the private ownership of real, usable property, and more and more upon property that is institutional and abstract, beyond individual control, such as money, insurance policies, certificates of deposit, stocks, and shares. And as our private economies become more abstract, the mutual, free helps and pleasures of family and community life will be supplanted by a kind of displaced or placeless citizenship and by commerce with impersonal and self-interested suppliers.

All of us, in fact, are now involved in destructive work or destructive pleasure or both. All of us are now directly dependent, economically and politically, upon the minds and ideas of people whom we do not know. Most of us have no way of knowing except, too late, by scandal or disaster, what is going on in the governments and the corporations. Most of us own no usable property. Most of us are watching the dispersal or disintegration of our families and communities.

Thus, although we are not slaves in name, and cannot be carried to market and sold as somebody else's legal chattels, we are free only within narrow limits. For all our talk about liberation and personal autonomy, there are few choices that we are free to make. What would be the point, for example, if a majority of our people decided to be self-employed?

The great enemy of freedom is the alignment of political power with wealth. This alignment destroys the commonwealth—that is, the natural wealth of localities and the local economies of household, neighborhood, and community—and so destroys democracy, of which the commonwealth is the foundation and the practical means. This happens—it is happening—because the alignment of wealth and power permits economic value to overturn value of any other kind. The value of everything is reduced to its market price. A thing not marketable has no value. It is increasingly apparent that we cannot value things except by selling them, and that we think it acceptable, and indeed respectable, to sell anything. For a number of years now the ruling political idea in my home state has been "Sell Kentucky." We speak more and more easily, too, of "selling" ideas. It is harder all the time to affirm the existence, or the right to exist, of a thing or an idea that "won't sell." Indeed, it is increasingly evident that we have replaced the old market on which people were sold with a new market on which people sell themselves.

Several months ago I attended the commencement exercises of a California university at which the graduates of the school of business wore "For Sale" signs around their necks. It was done as a joke, of course, a display of youthful high spirits, and yet it was inescapably a cynical joke, of the sort by which an embarrassing truth is flaunted. For, in fact, these graduates were for sale, they knew that they were, and they intended to be. They had just spent four years at a university to increase their "marketability." That some of the young women in the group undoubtedly were feminists only made the joke more cynical. But what most astonished and alarmed me was that a number of these graduates for sale were black. Had their forebears served and suffered and struggled in America for 368 years in order for these now certified and privileged few to sell themselves? Did they not know that only 122 years, two lifetimes, ago, their forebears had worn in effect that very sign? It seemed to me that I was witnessing the tragedy of history that the forgetfulness of history always is—and a tragedy not for blacks alone. If the people among us who know best, or ought to, what it means to be sold come to forget it or ignore it so far as to sell themselves, what is to become of the rest of us? If *they* have not learned better, how will the rest of us learn?

How, remembering their history, could those black graduates have worn those signs? Only, I think, by assuming, in very dangerous innocence, that their graduation into privilege exempted them from history. The danger is that there is no safety, no *dependable* safety, in privilege that is founded on greed, ignorance, and waste. And these people, after all, will remain black. What sign will they wear besides their expensive suits by which the police can tell them from their unemployed and unemployable brothers and sisters of the inner city?

* * *

There is no safety in belonging to the select few, for minority people or anybody else. If we are looking for insurance against want and oppression, we will find it only in our neighbors' prosperity and goodwill and, beyond that, in the good health of our worldly places, our homelands. If we were sincerely looking for a place of safety, for real security and success, then we would begin to turn to our communities—and not the communities simply of our human neighbors, but also of the water, earth, and air, the plants and animals, all the creatures with whom our local life is shared. We would be looking too for another kind of freedom. Our present idea of freedom is only the freedom to do as we please: to sell ourselves for a high salary, a home in the suburbs, and idle weekends. But that is a freedom dependent upon affluence, which is in turn dependent upon the rapid consumption of exhaustible supplies. The other kind of freedom is the freedom to take care of ourselves and of each other. The freedom of affluence opposes and contradicts the freedom of community life.

Our place of safety can only be the community, and not just one community, but many of them everywhere. Upon that depends all that we still claim to value: freedom, dignity, health, mutual help and affection, undestructive pleasure, and the rest. Human life, as most of us still would like to define it, is community life. And this brings me to the fourth problem, and the last, that I want to consider: How can "integration" be achieved—and what can it mean—in communities that are conspicuously disintegrating?

In a "Face to Face" television interview on July 8, 1967, Roy Wilkins described this problem pointedly enough:

> In the South we still have a great deal of Negro family stability and control, and community control of families, and the imposition of standards of conduct. In the North, with its great anonymous cities, Negro families come there, some-

> times they're disintegrated—but even where they're not, they are lost in a huge population; the minister doesn't keep tabs on them like the minister did in the small town back home; they don't know the police chief, and they don't care, and they don't know the judge and they don't know the things and the controls that operated in their home community. And when they come to Harlem, they're just "John Smith," and they can do as they please. They don't have to pay any attention to Mom and Pop, and the minister and the neighbor, or anybody who knows about them and helps to control them. So they run wild, some of them; others busy themselves going to night school and doing all the things that other people do.

With a few small differences, that describes a tragedy that is as white as it is black. The gravity of the problem is suggested by the fact that the disintegration Wilkins described coincides neatly with what a great many people, black and white, understand as freedom. People who wish to be free to pay no attention to anybody who knows them are not going to accept the constraints or pursue the freedoms of community life. They accept disintegration as the price or the sign of "success." Like the for-sale graduates of the California university, they think that mockery of history is "sophistication."

Mostly, we do not speak of our society as disintegrating. We would prefer not to call what we are experiencing social disintegration. But we are endlessly preoccupied with the symptoms: divorce, venereal disease, murder, rape, debt, bankruptcy, pornography, teenage pregnancy, fatherless children, motherless children, child suicide, public child-care, retirement homes, nursing homes, toxic waste, soil loss, soil and water and air pollution, government secrecy, government lying, government crime, civil violence, drug abuse, sexual promiscuity, abortion as "birth control," the explosion of garbage, hopeless poverty, unemployment, unearned wealth. We know the symptoms well enough. All the plagues of our time are symptoms of a general disintegration.

We are capable, really, only of the forcible integration of centralization—economic, political, military, and educational—and always at the cost of social and cultural disintegration. Our aim, it would appear, is to "integrate" ourselves into a limitless military-industrial city in which we all will be lost, and so may do as we please in the freedom either to run wild until we are caught or killed, or to do "all the things that other people do."

That we prefer to deal piecemeal with the problems of disintegration keeps them "newsworthy" and profitable to the sellers of cures. To see them as merely the symptoms of a greater problem would require hard thought, a change of heart, and a search for the fundamental causes.

Drug abuse, for example, will remain an easy political cause, and a lucrative business for everybody but the victims, so long as we take refuge in our meaningless distinction between legal and illegal drugs. How can we hope to stop the distribution of drugs in "the drug world" so long as we are unconcerned about the distribution of drugs by the drugstore? In fact, people use drugs, legal and illegal, because their lives are intolerably painful or dull. They hate their work and find no rest in their leisure. They are estranged from their families and their neighbors. It should tell us something that in healthy societies drug use is celebrative, convivial, and occasional, whereas among us it is lonely, shameful, and addictive. We need drugs, apparently, because we have lost each other.

* * *

But surely the most poignant symptom of disintegration is "integration" by the forced busing of schoolchildren to "achieve racial balance in the schools." This is an extremely risky subject for a white person to talk about because busing has become the major tool of integration, and so has taken on great force as a political symbol. Only a racist, it is assumed, can oppose busing for racial balance.

There are, nevertheless, some things about this practice that are wrong. For one thing, by focusing exclusively on the issue of racial balance in the schools, busing tends to distract attention from the much more widespread phenomenon of segregation by economic subdivision. I mean the division of urban and suburban economic classes—poor, blue collar, professional, very rich—into separate ghettos or enclaves. And this economic division is fractioned even more by the tendency of professional people and intellectuals to cohere in widely dispersed "networks," often to the virtual exclusion of community ties. Many people now feel more at home, and more at ease socially, at a professional convention than in the streets of their own neighborhood. But as the "successful" abandon the communities that they once shared with the unsuccessful, they forget the unsuccessful and leave them without examples or defenders. The children of the unsuccessful then have no models, or they have models only of the worst kind.

There are reasons for this economic segregation or disintegration, of course, and chief among them are economic and institutional centralization—and the automobile and TV, which are the technology of centralization. People don't work or shop or amuse themselves or go to church or school in their own neighborhoods anymore, and are therefore free to separate themselves from their workplaces and economic sources, and to sort themselves into economic categories in which, having no need for each other, they remain strangers. I assume that this is bad because I assume that it is good for people to know each other. I assume, especially, that it is good for people to know each other across the lines of economy and vocation. Professional people should know their clients outside their offices. Teachers should know the families of their students. University professors and intellectuals should know the communities and the households that will be affected by their ideas. Rich people and poor people should know each other. If this familiar knowledge does not exist, then these various groups will think of each other and deal with each other on the basis of stereotypes as vicious and ultimately as dangerous as the stereotypes of race.

It is in connection with this larger disintegration that busing for integration must be thought about, for busing (for school consolidation) was a tool of the larger disintegration long before it became a tool of racial integration.

My own children were bused to school from the first grade on. Their daily bus ride to and from school took about two hours of every day. This meant that they were under school discipline—expected to sit still, etc.—about a third again as long each day as their schoolmates in town. It also meant that they were under home discipline two hours a day less than the town children; they had that much less time for chores, homework, and free play. In my opinion, all this bus travel was damaging to the lives of my children both at school and at home. Moreover, the grade school that my children attended was nine miles, and their middle and high schools twelve miles, from home, well beyond the range of close or easy parental involvement. School consolidation thus involves a great expense of time and money that might be better spent in the education and upbringing of children.

Thus the question that I have had to ask myself out of my own experience is this: How can I be for busing as a tool of integration when I am against it as a tool of consolidation? My own experience suggests to me that busing for any reason is, in reality, a tool of disintegration. I believe in neighborhood schools for the same reasons that I believe in neighborhood shops and stores, for the same reasons that I believe in neighborhood.

There can be no greater blow to the integrity of a community than the loss of its school or loss of control of its school—which always means loss of control of its children. The breakdown of discipline and academic standards in the schools can only originate in, and can only cause, the breakdown of community life. The public school, separated from the community by busing (for whatever reason), government control, consolidation, and other "advances," has become a no-man's-land, a place existing in reference only to itself and to a theoretical "tomorrow's world." Neither teachers nor students feel themselves answerable to the community, for the school does not exist to serve the community. It exists to aid and abet the student's escape from the community into "tomorrow's world," in which community standards, it goes without saying, will not apply. The teachers are divided from the community by the shibboleths of "professional training," "professional standards," and "academic freedom." The students are divided from the community by the distance of school from home, by parental indifference to the affairs of a distant school not under their influence or control, and by changes in curriculum or teaching methods that make it impossible for many of them to get help from their elders. Teachers who are preparing students for jobs in "tomorrow's world," without reference to the local community or community anywhere, need not be surprised if their efforts are not enthusiastically affirmed by parents who are, after all, living in today's world.

The longer these imposed-from-above "solutions" continue, the more unsatisfactory they will prove to be. It is impossible to believe that people can be changed fundamentally by government requirement. People do not pay taxes voluntarily, for example, and they will not learn to do so in a thousand years of involuntary taxpaying. The only thing that a government requirement assures is a prolongation of government supervision. It is certain that the government should forbid racial injustice to the same extent that it should forbid injustice of any other kind. But that interracial liking and harmony can be the result of a government program is extremely doubtful. One may reasonably suspect, indeed, that government programs of social amelioration, such as welfare and busing, exist as poor apologies for the government's espousal of the economic and technological determinism that has virtually destroyed community life and community economy everywhere in the country.

A true and appropriate answer to our race problem, as to many others, would be a restoration of our communities — it being understood that a community, properly speaking, cannot exclude or mistreat any of its members. This is what we forgot during slavery and the industrialization that followed, and have never remembered. A proper community, we should remember also, is a commonwealth: a place, a resource, and an economy. It answers the needs, practical as well as social and spiritual, of its members — among them the need to need one another. The answer to the present alignment of political power with wealth is the restoration of the identity of community and economy.

Is this something that the government could help with? Of course it is.

Community cannot be made by government prescription and mandate, but the government, in its proper role as promoter of the general welfare, preserver of the public peace, and forbidder of injustice, could do much to promote the improvement of communities. If it wanted to, it could end its collusion with the wealthy and the corporations and the "special interests." It could stand, as it is supposed to, between wealth and power. It could assure the possibility that a poor person might hold office. It could protect, by strict forbiddings, the disruption of the integrity of a community or a local economy or an ecosystem by any sort of commercial or industrial enterprise, that is, it could enforce proprieties of scale. It could understand that economic justice does not consist in giving the most power to the most money.

The government *could* do such things. But we know well that it is not going to do them; it is not even going to consider doing them, because community integrity, and the decentralization of power and economy that it implies, is antithetical to the ambitions of the corporations. The government's aim, therefore, is racial indifference, not integrated communities. Does this mean that our predicament is hopeless? No. It only means that our predicament is extremely unfavorable, as the human predicament has often been.

What the government will or will not do is finally beside the point. If people do not have the government they want, then they will have a government that they must either change or endure. Finally, all the issues that I have discussed here are neither political nor economic, but moral and spiritual. What is at issue is our character as a people. It is necessary to look beyond the government to the possibility—one that seems to be growing—that people will reject what have been the prevailing assumptions, and begin to strengthen and defend their communities on their own.

We must be aware too of the certainty that the present way of things will eventually fail. If it fails quickly, by any of several predicted causes, then we will have no need, being absent, to worry about what to do next. If it fails slowly, and if we have been careful to preserve the most necessary and valuable things, then it may fail into a restoration of community life—that is, into understanding of our need to help and comfort each other.

Port Royal, Kentucky
Summer 1988

Feminism, the Body, and the Machine

Some time ago *Harper's* reprinted a short essay of mine in which I gave some of my reasons for refusing to buy a computer. Until that time, the vast numbers of people who disagree with my writings had mostly ignored them. An unusual number of people, however, neglected to ignore my insensitivity to the wonders of computer enhancement. Some of us, it seems, would be better off if we would just realize that this is already the best of all possible worlds, and is going to get even better if we will just buy the right equipment.

Harper's published only five of the letters the editors received in response to my essay, and they published only negative letters. But of the twenty letters received by the *Harper's* editors, who forwarded copies to me, three were favorable. This I look upon as extremely gratifying. If these letters may be taken as a fair sample, then one in seven of *Harper's* readers agreed with me. If I had guessed beforehand, I would have guessed that my supporters would have been fewer than one in a thousand. And so I suppose, after further reflection, that my surprise at the intensity of the attacks on me is mistaken. There are more of us than I thought. Maybe there is even a "significant number" of us.

Only one of the negative letters seemed to me to have much intelligence in it. That one was from R. N. Neff of Arlington, Virginia, who scored a direct hit: "Not to be obtuse, but being willing to bare my illiterate soul for all to see, is there indeed a 'work demonstrably better than Dante's' . . . which was written on a Royal standard typewriter?" I like this retort so well that I am tempted to count it a favorable response, raising the total to four. The rest of the negative replies, like the five published ones, were more feeling than intelligent. Some of them, indeed, might be fairly described as exclamatory.

One of the letter writers described me as "a fool" and "doubly a fool," but fortunately misspelled my name, leaving me a speck of hope that I am not the "Wendell Barry" he was talking about. Two others accused me of self-righteousness, by which they seem to have meant that they think they are righter than I think I am. And another accused me of being more concerned about my own moral purity than with "any ecological effect," thereby making the sort of razor-sharp philosophical distinction that could cause a person to be elected president.

But most of my attackers deal in feelings either feminist or technological, or both. The feelings expressed seem to be representative of what the state of public feeling currently permits to be felt, and of what public rhetoric currently permits to be said. The feelings, that is, are similar enough, from one letter to another, to be thought representative, and as representative letters they have an interest greater than the quarrel that occasioned them.

* * *

Without exception, the feminist letters accuse me of exploiting my wife, and they do not scruple to allow the most insulting implications of their indictment to fall upon my wife. They fail entirely to see that my essay does not give any support to their accusation—or if they see it, they do not care. My essay, in fact, does not characterize my wife beyond saying that she types my manuscripts and tells me what she thinks about them. It does not say what her motives are, how much work she does, or whether or how she is paid. Aside from saying that she is my wife and that I value the help she gives me with my work, it says nothing about our marriage. It says nothing about our economy.

There is no way, then, to escape the conclusion that my wife and I are subjected in these letters to a condemnation by category. My offense is that I am a man who receives some help from his wife; my wife's offense is that she is a woman who does some work for her husband—which work, according to her critics and mine, makes her a drudge, exploited by a conventional subservience. And my detractors have, as I say, no evidence to support any of this. Their accusation rests on a syllogism of the flimsiest sort: my wife helps me in my work, some wives who have helped their husbands in their work have been exploited, therefore my wife is exploited.

This, of course, outrages justice to about the same extent that it insults intelligence. Any respectable system of justice exists in part as a protection against such accusations. In a just society nobody is expected to plead guilty to a general indictment, because in a just society nobody can be convicted on a general indictment. What is required for a just conviction is a particular accusation that can be *proved*. My accusers have made no such accusation against me.

* * *

That feminists or any other advocates of human liberty and dignity should resort to insult and injustice is regrettable. It is equally regrettable that all of the feminist attacks on my essay implicitly deny the validity of two decent and probably necessary possibilities: marriage as a state of mutual help, and the household as an economy.

Marriage, in what is evidently its most popular version, is now on the one hand an intimate "relationship" involving (ideally) two successful careerists in the same bed, and on the other hand a sort of private political system in which rights and interests must be constantly asserted and defended. Marriage, in other words, has now taken the form of divorce: a prolonged and impassioned negotiation as to how things shall be divided. During their understandably temporary association, the "married" couple will typically consume a large quantity of merchandise and a large portion of each other.

The modern household is the place where the consumptive couple do their consuming. Nothing productive is done there. Such work as is done there is done at the expense of the resident couple or family, and to the profit of suppliers of energy and household technology. For entertainment, the inmates consume television or purchase other consumable diversion elsewhere.

There are, however, still some married couples who understand themselves as belonging to their marriage, to each other, and to their children. What they have they have in common, and so, to them, helping each other does not seem merely to damage their ability to compete against each other. To them, "mine" is not so powerful or necessary a pronoun as "ours."

This sort of marriage usually has at its heart a household that is to some extent productive. The couple, that is, makes around itself a household economy that involves the work of both wife and husband, that gives them a measure of economic independence and self-protection, a measure of self-employment, a measure of freedom, as well as a common ground and a common satisfaction. Such a household economy may employ the disciplines and skills of housewifery, of carpentry and other trades of building and maintenance, of gardening and other branches of subsistence agriculture, and even of woodlot management and wood-cutting. It may also involve a "cottage industry" of some kind, such as a small literary enterprise.

It is obvious how much skill and industry either partner may put into such a household and what a good economic result such work may have, and yet it is a kind of work now frequently held in contempt. Men in general were the first to hold it in contempt as they departed from it for the sake of the professional salary or the hourly wage, and now it is held in contempt by such feminists as those who attacked my essay. Thus farm wives who help to run the kind of household economy that I have described are apt to be asked by feminists, and with great condescension, "But what do you *do*?" By this they invariably mean that there is something better to do than to make one's marriage and household, and by better they invariably mean "employment outside the home."

* * *

I know that I am in dangerous territory, and so I had better be plain: what I have to say about marriage and household I mean to apply to men as much as to women. I do not believe that there is anything better to do than to make one's marriage and household, whether one is a man or a woman. I do not believe that "employment outside the home" is as valuable or important or satisfying as employment at home, for either men or women. It is clear to me from my experience as a teacher, for example, that children need an ordinary daily association with *both* parents. They need to see their parents at work; they need, at first, to play at the work they see their parents doing, and then they need to work with their parents. It does not matter so much that this working together should be what is called "quality time," but it matters a great deal that the work done should have the dignity of economic value.

I should say too that I understand how fortunate I have been in being able to do an appreciable part of my work at home. I know that in many marriages both husband and wife are now finding it necessary to work away from home. This issue, of course, is troubled by the question of what is meant by "necessary," but it is true that a family living that not so long ago was ordinarily supplied by one job now routinely requires two or more. My interest is not to quarrel with individuals, men or women, who work away from home, but rather to ask why we should consider this general working away from home to be a desirable state of things, either for people or for marriage, for our society or for our country.

If I had written in my essay that my wife worked as a typist and editor for a publisher, doing the same work that she does for me, no feminists, I daresay, would have written to *Harper's* to attack me for exploiting her—even though, for all they knew, I might have forced her to do such work in order to keep me in gambling money. It would have been assumed as a matter of course that if she had a job away from home she was a "liberated woman," possessed of a dignity that no home could confer upon her.

As I have said before, I understand that one cannot construct an adequate public defense of a private life. Anything that I might say here about my marriage would be immediately (and rightly) suspect on the ground that it would be only *my* testimony. But for the sake of argument, let us suppose that whatever work my wife does, as a member of our marriage and household, she does both as a full economic partner and as her own boss, and let us suppose that the economy we have is adequate to our needs. Why, granting that supposition, should anyone assume that my wife would increase her freedom or dignity or satisfaction by becoming the employee of a boss, who would be in turn also a corporate underling and in no sense a partner?

Why would any woman who would refuse, properly, to take the marital vow of obedience (on the ground, presumably, that subservience to a mere human being is beneath human dignity) then regard as "liberating" a job that puts her under the authority of a boss (man or woman) whose authority specifically requires and expects obedience? It is easy enough to see why women came to object to the role of Blondie, a mostly decorative custodian of a degraded, consumptive modern household, preoccupied with clothes, shopping, gossip, and outwitting her husband. But are we to assume that one may fittingly cease to be Blondie by becoming Dagwood? Is the life of a corporate underling—even acknowledging that corporate underlings are well paid—an acceptable end to our quest for human dignity and worth? It is clear enough by now that one does not cease to be an underling by reaching "the top." Corporate life is composed only of lower underlings and higher underlings. Bosses are everywhere, and all the bosses are underlings. This is invariably revealed when the time comes for accepting responsibility for something unpleasant, such as the

Exxon fiasco in Prince William Sound, for which certain lower underlings are blamed but no higher underling is responsible. The underlings at the top, like telephone operators, have authority and power, but no responsibility.

And the oppressiveness of some of this office work defies belief. Edward Mendelson (in the *New Republic*, February 22, 1988) speaks of "the office worker whose computer keystrokes are monitored by the central computer in the personnel office, and who will be fired if the keystrokes-per-minute figure doesn't match the corporate quota." (Mr. Mendelson does not say what form of drudgery this worker is being saved from.) And what are we to say of the diversely skilled country housewife who now bores the same six holes day after day on an assembly line? What higher form of womanhood or humanity is she evolving toward?

How, I am asking, can women improve themselves by submitting to the same specialization, degradation, trivialization, and tyrannization of work that men have submitted to? And that question is made legitimate by another: How have men improved themselves by submitting to it? The answer is that men have not, and women cannot, improve themselves by submitting to it.

Women have complained, justly, about the behavior of "macho" men. But despite their he-man pretensions and their captivation by masculine heroes of sports, war, and the Old West, most men are now entirely accustomed to obeying and currying the favor of their bosses. Because of this, of course, they hate their jobs—they mutter, "Thank God it's Friday" and "Pretty good for Monday"—but they do as they are told. They are more compliant than most housewives have been. Their characters combine feudal submissiveness with modern helplessness. They have accepted almost without protest, and often with relief, their dispossession of any usable property and, with that, their loss of economic independence and their consequent subordination to bosses. They have submitted to the destruction of the household economy and thus of the household, to the loss of home employment and self-employment, to the disintegration of their families and communities, to the desecration and pillage of their country, and they have continued abjectly to believe, obey, and vote for the people who have most eagerly abetted this ruin and who have most profited from it. These men, moreover, are helpless to do anything for themselves or anyone else without money, and so for money they do whatever they are told. They know that their ability to be useful is precisely defined by their willingness to be somebody else's tool. Is it any wonder that they talk tough and worship athletes and cowboys? Is it any wonder that some of them are violent?

It is clear that women cannot justly be excluded from the daily fracas by which the industrial economy divides the spoils of society and nature, but their inclusion is a poor justice and no reason for applause. The enterprise is as devastating with women in it as it was before. There is no sign that women are exerting a "civilizing influence" upon it. To have an equal part in our juggernaut of national vandalism is to be a vandal. To call this vandalism "liberation" is to prolong, and even ratify, a dangerous confusion that was once principally masculine.

A broader, deeper criticism is necessary. The problem is not just the exploitation of women by men. A greater problem is that women and men alike are consenting to an economy that exploits women and men and everything else.

Another decent possibility my critics implicitly deny is that of work as a gift. Not one of them supposed that my wife may be a consulting engineer who helps me in her spare time out of the goodness of her heart; instead they suppose that she is "a household drudge." But what appears to infuriate them the most is their supposition that she works for nothing. They assume—and this is the orthodox assumption of the industrial economy—that the only help worth giving is not given at all, but sold. Love, friendship, neighborliness, compassion, duty—what are they? We are realists. We will be most happy to receive your check.

* * *

The various reductions I have been describing are fairly directly the results of the ongoing revolution of applied science known as "technological progress." This revolution has provided the means by which both the productive and the consumptive capacities of people could be detached from household and community and made to serve other people's purely economic ends. It has provided as well a glamor of newness, ease, and affluence that made it seductive even to those who suffered most from it. In its more recent history especially, this revolution has been successful in putting unheard-of quantities of consumer goods and services within the reach of ordinary people. But the technical means of this popular "affluence" has at the same time made possible the gathering of the real property and the real power of the country into fewer and fewer hands.

Some people would like to think that this long sequence of industrial innovations has changed human life and even human nature in fundamental ways. Perhaps it has—but, arguably, almost always for the worse. I know that "technological progress" can be defended, but I observe that the defenses are invariably quantitative—catalogs of statistics on the ownership of automobiles and television sets, for example, or on the increase of life expectancy—and I see that these statistics are always kept carefully apart from the related statistics of soil loss, pollution, social disintegration, and so forth. That is to say, there is never an effort to determine the *net* result of this progress. The voice of its defenders is not that of the responsible bookkeeper, but that of the propagandist or salesman, who says that the net gain is more than 100 percent—that the thing we have bought has perfectly replaced everything it has cost, and added a great deal more: "You just can't lose!" We thus have got rich by spending, just as the advertisers have told us we would, and the best of all possible worlds is getting better every day.

The statistics of life expectancy are favorites of the industrial apologists, because they are perhaps the hardest to argue with. Nevertheless, this emphasis on longevity is an excellent example of the way the isolated aims of the industrial mind reduce and distort human life, and also the way statistics corrupt the truth. A long life has indeed always been thought desirable; everything that is alive apparently wishes to continue to live. But until our own time, that sentence would have been qualified: long life is desirable and everything wishes to live *up to a point*. Past a certain point, and in certain conditions, death becomes preferable to life. Moreover, it was generally agreed that a good life was preferable to one that was merely long, and that the goodness of a life could not be determined by its length. The statisticians of longevity ignore good in both its senses; they do not ask if the prolonged life is virtuous, or if it is satisfactory. If the life is that of a vicious criminal, or if it is inched out in a veritable hell of captivity within the medical industry, no matter—both become statistics to "prove" the good luck of living in our time.

But in general, apart from its own highly specialized standards of quantity and efficiency, "technological progress" has produced a social and ecological decline. Industrial war, except by the most fanatically narrow standards, is worse than war used to be. Industrial agriculture, except by the standards of quantity and mechanical efficiency, diminishes everything it affects. Industrial workmanship is certainly worse than traditional workmanship, and is getting shoddier every day. After forty-odd years, the evidence is everywhere that television, far from proving a great tool of education, is a tool of stupefaction and disintegration. Industrial education has abandoned the old duty of passing on the cultural and intellectual inheritance in favor of baby-sitting and career preparation.

After several generations of "technological progress," in fact, we have become a people who *cannot* think about anything important. How far down in the natural order do we have to go to find creatures who raise their young as indifferently as industrial humans now do? Even the English sparrows do not let loose into the streets young sparrows who have no notion of their identity or their adult responsibilities. When else in history would you find "educated" people who know more about sports than about the history of their country, or uneducated people who do not know the stories of their families and communities?

* * *

To ask a still more obvious question, what is the purpose of this technological progress? What higher aim do we think it is serving? Surely the aim cannot be the integrity or happiness of our families, which we have made subordinate to the education system, the television industry, and the consumer economy. Surely it cannot be the integrity or health of our communities, which we esteem even less than we esteem our families. Surely it cannot be love of our country, for we are far more concerned about the desecration of the flag than we are about the desecration of our land. Surely it cannot be the love of God, which counts for at least as little in the daily order of business as the love of family, community, and country.

The higher aims of "technological progress" are money and ease. And this exalted greed for money and ease is disguised and justified by an obscure, cultish faith in "the future." We do as we do, we say, "for the sake of the future" or "to make a better future for our children." How we can hope to make a good future by doing badly in the present, we do not say. We cannot think about the future, of course, for the future does not exist: the existence of the future is an article of faith. We can be assured only that, if there is to be a future, the good of it is already implicit in the good things of the present. We do not need to plan or devise a "world of the future"; if we take care of the world of the present, the future will have received full justice from us. A good future is implicit in the soils, forests, grasslands, marshes, deserts, mountains, rivers, lakes, and oceans that we have now, and in the good things of human culture that we have now; the only valid "futurology" available to us is to take care of those things. We have no need to contrive and dabble at "the future of the human race"; we have the same pressing need that we have always had—to love, care for, and teach our children.

And so the question of the desirability of adopting any technological innovation is a question with two possible answers—not one, as has been commonly assumed. If one's motives are money, ease, and haste to arrive in a technologically determined future, then the answer is foregone, and there is, in fact, no question, and no thought. If one's motive is the love of family, community, country, and God, then one will have to think, and one may have to decide that the proposed innovation is undesirable.

The question of how to end or reduce dependence on some of the technological innovations already adopted is a baffling one. At least, it baffles me. I have not been able to see, for example, how people living in the country, where there is no public transportation, can give up their automobiles without becoming less useful to each other. And this is because, owing largely to the influence of the automobile, we live too far from each other, and from the things we need, to be able to get about by any other means. Of course, you *could* do without an automobile, but to do so you would have to disconnect yourself from many obligations. Nothing I have so far been able to think about this problem has satisfied me.

But if we have paid attention to the influence of the automobile on country communities, we know that the desirability of technological innovation is an issue that requires thinking about, and we should have acquired some ability to think about it. Thus if I am partly a writer, and I am offered an expensive machine to help me write, I ought to ask whether or not such a machine is desirable.

I should ask, in the first place, whether or not I wish to purchase a solution to a problem that I do not have. I acknowledge that, as a writer, I need a lot of help. And I have received an abundance of the best of help from my wife, from other members of my family, from friends, from teachers, from editors, and sometimes from readers. These people have helped me out of love or friendship, and perhaps in exchange for some help that I have given them. I suppose I should leave open the possibility that I need more help than I am getting, but I would certainly be ungrateful and greedy to think so.

But a computer, I am told, offers a kind of help that you can't get from other humans; a computer will help you to write faster, easier, and more. For a while, it seemed to me that every university professor I met told me this. Do I, then, want to write faster, easier, and more? No. My standards are not speed, ease, and quantity. I have already left behind too much evidence that, writing with a pencil, I have written too fast, too easily, and too much. I would like to be a *better* writer, and for that I need help from other humans, not a machine.

The professors who recommended speed, ease, and quantity to me were, of course, quoting the standards of their universities. The chief concern of the industrial system, which is to say the present university system, is to cheapen work by increasing volume. But implicit in the professors' recommendation was the idea that one needs to be up with the times. The pace-setting academic intellectuals have lately had a great hankering to be up with the times. They don't worry about keeping up with the Joneses: as intellectuals, they know that they are supposed to be Nonconformists and Independent Thinkers living at the Cutting Edge of Human Thought. And so they are all a-dither to keep up with the times—which means adopting the latest technological innovations as soon as the Joneses do.

Do I wish to keep up with the times? No.

* * *

My wish simply is to live my life as fully as I can. In both our work and our leisure, I think, we should be so employed. And in our time this means that we must save ourselves from the products that we are asked to buy in order, ultimately, to replace ourselves.

The danger most immediately to be feared in "technological progress" is the degradation and obsolescence of the body. Implicit in the technological revolution from the beginning has been a new version of an old dualism, one always destructive, and now more destructive than ever. For many centuries there have been people who looked upon the body, as upon the natural world, as an encumbrance of the soul, and so have hated the body, as they have hated the natural world, and longed to be free of it. They have seen the body as intolerably imperfect by spiritual standards. More recently, since the beginning of the technological revolution, more and more people have looked upon the body, along with the rest of the natural creation, as intolerably imperfect by mechanical standards. They see the body as an encumbrance of the mind—the mind, that is, as reduced to a set of mechanical ideas that can be implemented in machines—and so they hate it and long to be free of it. The body has limits that the machine does not have; therefore, remove the body from the machine so that the machine can continue as an unlimited idea.

It is odd that simply because of its "sexual freedom" our time should be considered extraordinarily physical. In fact, our "sexual revolution" is mostly an industrial phenomenon, in which the body is used as an idea of pleasure or a pleasure machine with the aim of "freeing" natural pleasure from natural consequence. Like any other industrial enterprise, industrial sexuality seeks to conquer nature by exploiting it and ignoring the consequences, by denying any connection between nature and spirit or body and soul, and by evading social responsibility. The spiritual, physical, and economic costs of this "freedom" are immense, and are characteristically belittled or ignored. The diseases of sexual irresponsibility are regarded as a technological problem and an affront to liberty. Industrial sex, characteristically, establishes its freeness and goodness by an industrial accounting, dutifully toting up numbers of "sexual partners," orgasms, and so on, with the inevitable industrial implication that the body is somehow a limit on the idea of sex, which will be a great deal more abundant as soon as it can be done by robots.

This hatred of the body and of the body's life in the natural world, always inherent in the technological revolution (and sometimes explicitly and vengefully so), is of concern to an artist because art, like sexual love, is of the body. Like sexual love, art is of the mind and spirit also, but it is made with the body and it appeals to the senses. To reduce or shortcut the intimacy of the body's involvement in the making of a work of art (that is, of any artifice, anything made by art) inevitably risks reducing the work of art and the art itself. In addition to the reasons I gave previously, which I still believe are good reasons, I am not going to use a computer because I don't want to diminish or distort my bodily involvement in my work. I don't want to deny myself the *pleasure* of bodily involvement in my work, for that pleasure seems to me to be the sign of an indispensable integrity.

At first glance, writing may seem not nearly so much an art of the body as, say, dancing or gardening or carpentry. And yet language is the most intimately physical of all the artistic means. We have it palpably in our mouths; it is our *langue*, our tongue. Writing it, we shape it with our hands. Reading aloud what we have written—as we must do, if we are writing carefully—our language passes in at the eyes, out at the mouth, in at the ears; the words are immersed and steeped in the senses of the body before they make sense in the mind. They *cannot* make sense in the mind until they have made sense in the body. Does shaping one's words with one's own hand impart character and quality to them, as does speaking them with one's own tongue to the satisfaction of one's own ear? There is no way to prove that it does. On the other hand, there is no way to prove that it does not, and I believe that it does.

The act of writing language down is not so insistently tangible an act as the act of building a house or playing the violin. But to the extent that it is tangible, I love the tangibility of it. The computer apologists, it seems to me, have greatly underrated the value of the handwritten manuscript as an artifact. I don't mean that a writer should be a fine calligrapher and write for exhibition, but rather that handwriting has a valuable influence on the work so written. I am certainly no calligrapher, but my handwritten pages have a homemade, handmade look to them that both pleases me in itself and suggests the possibility of ready correction. It looks hospitable to improvement. As the longhand is transformed into typescript and then into galley proofs and the printed page, it seems increasingly to resist improvement. More and more spunk is required to mar the clean, final-looking lines of type. I have the notion—again not provable—that the longer I keep a piece of work in longhand, the better it will be.

To me, also, there is a significant difference between ready correction and easy correction. Much is made of the ease of correction in computer work, owing to the insubstantiality of the light-image on the screen; one presses a button and the old version disappears, to be replaced by the new. But because of the substantiality of paper and the consequent difficulty involved, one does not handwrite or typewrite a new page every time a correction is made. A handwritten or typewritten page therefore is usually to some degree a palimpsest; it contains parts and relics of its own history—erasures, passages crossed out, interlineations—suggesting that there is something to go back to as well as something to go forward to. The light-text on the computer screen, by contrast, is an artifact typical of what can only be called the industrial present, a present absolute. A computer destroys the sense of historical succession, just as do other forms of mechanization. The well-crafted table or cabinet embodies the memory of (because it embodies respect for) the tree it was made of and the forest in which the tree stood. The work of certain potters embodies the memory that the clay was dug from the earth. Certain farms contain hospitably the remnants and reminders of the forest or prairie that preceded them. It is possible even for towns and cities to remember farms and forests or prairies. All good human work remembers its history. The best writing, even when printed, is full of intimations that it is the present version of earlier versions of itself, and that its maker inherited the work and the ways of earlier makers. It thus keeps, even in print, a suggestion of the quality of the handwritten page; it is a palimpsest.

Something of this undoubtedly carries over into industrial products. The plastic Clorox jug has a shape and a loop for the forefinger that recalls the stoneware jug that went before it. But something vital is missing. It embodies no memory of its source or sources in the earth or of any human hand involved in its shaping. Or look at a large factory or a power plant or an airport, and see if you can imagine—even if you know—what was there before. In such things the materials of the world have entered a kind of orphanhood.

It would be uncharitable and foolish of me to suggest that nothing good will ever be written on a computer. Some of my best friends have computers. I have only said that a computer cannot help you to write *better*, and I stand by that.

(In fact, I know a publisher who says that under the influence of computers—or of the immaculate copy that computers produce—many writers are now writing worse.) But I do say that in using computers writers are flirting with a radical separation of mind and body, the elimination of the work of the body from the work of the mind. The text on the computer screen, and the computer printout too, has a sterile, untouched, factorymade look, like that of a plastic whistle or a new car. The body does not do work like that. The body *characterizes* everything it touches. What it makes it traces over with the marks of its pulses and breathings, its excitements, hesitations, flaws, and mistakes. On its good work, it leaves the marks of skill, care, and love persisting through hesitations, flaws, and mistakes. And to those of us who love and honor the life of the body in this world, these marks are precious things, necessities of life.

But writing is of the body in yet another way. It is preeminently a walker's art. It can be done on foot and at large. The beauty of its traditional equipment is simplicity. And cheapness. Going off to the woods, I take a pencil and some paper (*any* paper—a small notebook, an old envelope, a piece of a feed sack), and I am as well equipped for my work as the president of IBM. I am also free, for the time being at least, of everything that IBM is hooked to. My thoughts will not be coming to me from the power structure or the power grid, but from another direction and way entirely. My mind is free to go with my feet.

I know that there are some people, perhaps many, to whom you cannot appeal on behalf of the body. To them, disembodiment is a goal, and they long for the realm of pure mind—or pure machine; the difference is negligible. Their departure from their bodies, obviously, is much to be desired, but the rest of us had better be warned: they are going to cause a lot of dangerous commotion on their way out.

* * *

Some of my critics were happy to say that my refusal to use a computer would not do any good. I have argued, and am convinced, that it will at least do *me* some good, and that it may involve me in the preservation of some cultural goods. But what they meant was real, practical, public good. They meant that the materials and energy I save by not buying a computer will not be "significant." They meant that no individual's restraint in the use of technology or energy will be "significant." That is true.

But each one of us, by "insignificant" individual abuse of the world, contributes to a general abuse that is devastating. And if I were one of thousands or millions of people who could afford a piece of equipment, even one for which they had a conceivable "need," and yet did not buy it, *that* would be "significant." Why, then, should I hesitate for even a moment to be one, even the first one, of that "significant" number? Thoreau gave the definitive reply to the folly of "significant numbers" a long time ago: Why should anybody wait to do what is right until everybody does it? It is not "significant" to love your own children or to eat your own dinner, either. But normal humans will not wait to love or eat until it is mandated by an act of Congress.

One of my correspondents asked where one is to draw the line. That question returns me to the bewilderment I mentioned earlier: I am unsure where the line ought to be drawn, or how to draw it. But it is an intelligent question, worth losing some sleep over.

I know how to draw the line only where it is easy to draw. It is easy—it is even a luxury—to deny oneself the use of a television set, and I zealously practice that form of self-denial. Every time I see television (at other people's houses), I am more inclined to congratulate myself on my deprivation. I have no doubt, as I have said, that I am better off without a computer. I joyfully deny myself a motorboat, a camping van, an off-road vehicle, and every other kind of recreational machinery. I have, and want, no "second home." I suffer very comfortably the lack of colas, TV dinners, and other counterfeit foods and beverages.

I am, however, still in bondage to the automobile industry and the energy companies, which have nothing to recommend them except our dependence on them. I still fly on airplanes, which have nothing to recommend them but speed; they are inconvenient, uncomfortable, undependable, ugly, stinky, and scary. I still cut my wood with a chainsaw, which has nothing to recommend it but speed, and has all the faults of an airplane, except it does not fly.

It is plain to me that the line ought to be drawn without fail wherever it can be drawn easily. And it ought to be easy (though many do not find it so) to refuse to buy what one does not need. If you are already solving your problem with the equipment you have—a pencil, say—why solve it with something more expensive and more damaging? If you don't have a problem, why pay for a solution? If you love the freedom and elegance of simple tools, why encumber yourself with something complicated?

And yet, if we are ever again to have a world fit and pleasant for little children, we are surely going to have to draw the line where it is *not* easily drawn. We are going to have to learn to give up things that we have learned (in only a few years, after all) to "need." I am not an optimist; I am afraid that I won't live long enough to escape my bondage to the machines. Nevertheless, on every day left to me I will search my mind and circumstances for the means of escape. And I am not without hope. I knew a man who, in the age of chainsaws, went right on cutting his wood with a handsaw and an axe. He was a healthier and a saner man than I am. I shall let his memory trouble my thoughts.

1989

Think Little

First there was Civil Rights, and then there was the War, and now it is the Environment. The first two of this sequence of causes have already risen to the top of the nation's consciousness and declined somewhat in a remarkably short time. I mention this in order to begin with what I believe to be a justifiable skepticism. For it seems to me that the Civil Rights Movement and the Peace Movement, as popular causes in the electronic age, have partaken far too much of the nature of fads. Not for all, certainly, but for too many they have been the fashionable politics of the moment. As causes they have been undertaken too much in ignorance; they have been too much simplified; they have been powered too much by impatience and guilt of conscience and short-term enthusiasm, and too little by an authentic social vision and long-term conviction and deliberation. For most people those causes have remained almost entirely abstract; there has been too little personal involvement, and too much involvement in organizations that were insisting that *other* organizations should do what was right.

There is considerable danger that the Environment Movement will have the same nature: that it will be a public cause, served by organizations that will self-righteously criticize and condemn other organizations, inflated for a while by a lot of public talk in the media, only to be replaced in its turn by another fashionable crisis. I hope that will not happen, and I believe that there are ways to keep it from happening, but I know that if this effort is carried on solely as a public cause, if millions of people cannot or will not undertake it as a *private* cause as well, then it is *sure* to happen. In five years the energy of our present concern will have petered out in a series of public gestures—and no doubt in a series of empty laws—and a great, and perhaps the last, human opportunity will have been lost.

It need not be that way. A better possibility is that the movement to preserve the environment will be seen to be, as I think it has to be, not a digression from the civil rights and peace movements, but the logical culmination of those movements. For I believe that the separation of these three problems is artificial. They have the same cause, and that is the mentality of greed and exploitation. The mentality that exploits and destroys the natural environment is the same that abuses racial and economic minorities, that imposes on young men the tyranny of the military draft, that makes war against peasants and women and children with the indifference of technology. The mentality that destroys a watershed and then panics at the threat of flood is the same mentality that gives institutionalized insult to black people and then panics at the prospect of race riots. It is the same mentality that can mount deliberate warfare against a civilian population and then express moral shock at the logical consequence of such warfare at My Lai. We would be fools to believe that we could solve any one of these problems without solving the others.

To me, one of the most important aspects of the environmental movement is that it brings us not just to another public crisis, but to a crisis of the protest movement itself. For the environmental crisis should make it dramatically clear, as perhaps it has not always been before, that there is no public crisis that is not also private. To most advocates of civil rights, racism has seemed mostly the fault of someone else. For most advocates of peace the war has been a remote reality, and the burden of the blame has seemed to rest mostly on the government. I am certain that these crises have been more private, and that we have each suffered more from them and been more responsible for them than has been readily apparent, but the connections have been difficult to see. Racism and militarism have been institutionalized among us for too long for our personal involvement in those evils to be easily apparent to us. Think, for example, of all the Northerners who assumed—until black people attempted to move into *their* neighborhoods—that racism was a Southern phenomenon. And think how quickly—one might almost say how naturally—among some of its members the peace movement has spawned policies of deliberate provocation and violence.

* * *

But the environmental crisis rises closer to home. Every time we draw a breath, every time we drink a glass of water, every time we eat a bite of food we are suffering from it. And more important, every time we indulge in, or depend on, the wastefulness of our economy—and our economy's first principle is waste—we are *causing* the crisis. Nearly every one of us, nearly every day of his life, is contributing *directly* to the ruin of this planet. A protest meeting on the issue of environmental abuse is not a convocation of accusers, it is a convocation of the guilty. That realization ought to clear the smog of self-righteousness that has almost conventionally hovered over these occasions, and let us see the work that is to be done.

In this crisis it is certain that every one of us has a public responsibility. We must not cease to bother the government and the other institutions to see that they never become comfortable with easy promises. For myself, I want to say that I hope never again to go to Frankfort to present a petition to the governor on an issue so vital as that of strip mining, only to be dealt with by some ignorant functionary—as several of us were not so long ago, the governor himself being "too busy" to receive us. Next time I will go prepared to wait as long as necessary to see that the petitioners' complaints and their arguments are heard *fully*—and by the governor. And then I will hope to find ways to keep those complaints and arguments from being forgotten until something is done to relieve them. The time is past when it was enough merely to elect our officials. We will have to elect them and then go and *watch* them and keep our hands on them, the way the coal companies do. We have made a tradition in Kentucky of putting self-servers, and worse, in charge of our vital interests. I am sick of it. And I think that one way to change it is to make Frankfort a less comfortable place. I believe in American political principles, and I will not sit idly by and see those principles destroyed by sorry practice. I am ashamed and deeply distressed that American government should have become the chief cause of disillusionment with American principles.

And so when the government in Frankfort again proves too stupid or too blind or too corrupt to see the plain truth and to act with simple decency, I intend to be there, and I trust that I won't be alone. I hope, moreover, to be there, not with a sign or a slogan or a button, but with the facts and the arguments. A crowd whose discontent has risen no higher than the level of slogans is *only* a crowd. But a crowd that understands the reasons for its discontent and knows the remedies is a vital community, and it will have to be reckoned with. I would rather go before the government with two people who have a competent understanding of an issue, and who therefore deserve a hearing, than with two thousand who are vaguely dissatisfied.

But even the most articulate public protest is not enough. We don't live in the government or in institutions or in our public utterances and acts, and the environmental crisis has its roots in our *lives*. By the same token, environmental health will also be rooted in our lives. That is, I take it, simply a fact, and in the light of it we can see how superficial and foolish we would be to think that we could correct what is wrong merely by tinkering with the institutional machinery. The changes that are required are fundamental changes in the way we are living.

* * *

What we are up against in this country, in any attempt to invoke private responsibility, is that we have nearly destroyed private life. Our people have given up their independence in return for the cheap seductions and the shoddy merchandise of so-called "affluence." We have delegated all our vital functions and responsibilities to salesmen and agents and bureaus and experts of all sorts. We cannot feed or clothe ourselves, or entertain ourselves, or communicate with each other, or be charitable or neighborly or loving, or even respect ourselves, without recourse to a merchant or a corporation or a public-service organization or an agency of the government or a style-setter or an expert. Most of us cannot think of dissenting from the opinions or the actions of one organization without first forming a new organization. Individualism is going around these days in uniform, handing out the party line on individualism. Dissenters want to publish their personal opinions over a thousand signatures.

The Confucian *Great Digest* says that the "chief way for the production of wealth" (and he is talking about real goods, not money) is "that the producers be many and that the mere consumers be few. . . ." But even in the much-publicized rebellion of the young against the materialism of the affluent society, the consumer mentality is too often still intact: the standards of behavior are still those of kind and quantity, the security sought is still the security of numbers, and the chief motive is still the consumer's anxiety that he is missing out on what is "in." In this state of total consumerism—which is to say a state of helpless dependence on things and services and ideas and motives that we have forgotten how to provide ourselves—all meaningful contact between ourselves and the earth is broken. We do not understand the earth in terms either of what it offers us or of what it requires of us, and I think it is the rule that people inevitably destroy what they do not understand. Most of us are not directly responsible for strip mining and extractive agriculture and other forms of environmental abuse. But we are guilty nevertheless, for we connive in them by our ignorance. We are ignorantly dependent on them. We do not know enough about them; we do not have a particular enough sense of their danger. Most of us, for example, not only do not know how to produce the best food in the best way—we don't know how to produce any kind in any way. Our model citizen is a sophisticate who before puberty understands how to produce a baby, but who at the age of thirty will not know how to produce a potato. And for this condition we have elaborate rationalizations, instructing us that dependence for everything on somebody else is efficient and economical and a scientific miracle. I say, instead, that it is madness, mass produced. A man who understands the weather only in terms of golf is participating in a chronic public insanity that either he or his descendants will be bound to realize as suffering. I believe that the death of the world is breeding in such minds much more certainly and much faster than in any political capital or atomic arsenal.

For an index of our loss of contact with the earth we need only look at the condition of the American farmer—who must in our society, as in every society, enact man's dependence on the land, and his responsibility to it. In an age of unparalleled affluence and leisure, the American farmer is harder pressed and harder worked than ever before; his margin of profit is small, his hours are long; his outlays for land and equipment and the expenses of maintenance and operation are growing rapidly greater; he cannot compete with industry for labor; he is being forced more and more to depend on the use of destructive chemicals and on the wasteful methods of haste and anxiety. As a class, farmers are one of the despised minorities. So far as I can see, farming is considered marginal or incidental to the economy of the country, and farmers, when they are thought of at all, are thought of as hicks and yokels whose lives do not fit into the modern scene. The average American farmer is now an old man whose sons have moved away to the cities. His knowledge, and his intimate connection with the land, are about to be lost. The small independent farmer is going the way of the small independent craftsmen and storekeepers. He is being forced off the land into the cities, his place taken by absentee owners, corporations, and machines. Some would justify all this in the name of efficiency. As I see it, it is an enormous social and economic and cultural blunder. For the small farmers who lived on their farms *cared* about their land. And given their established connection to their land—which was often hereditary and traditional as well as economic—they could have been encouraged to care for it more competently than they have so far. The corporations and machines that replace them will never be bound to the land by the sense of birthright and continuity, or by the love that enforces care. They will be bound by the rule of efficiency, which takes thought only of the volume of the year's produce, and takes no thought of the slow increment of the life of the land, not measurable in pounds or dollars, which will assure the livelihood and the health of the coming generations.

If we are to hope to correct our abuses of each other and of other races and of our land, and if our effort to correct these abuses is to be more than a political fad that will in the long run be only another form of abuse, then we are going to have to go far beyond public protest and political action. We are going to have to rebuild the substance and the integrity of private life in this country. We are going to have to gather up the fragments of knowledge and responsibility that we have parceled out to the bureaus and the corporations and the specialists, and we are going to have to put those fragments back together again in our own minds and in our families and households and neighborhoods. We need better government, no doubt about it. But we also need better minds, better friendships, better marriages, better communities. We need persons and households that do not have to wait upon organizations, but can make necessary changes in themselves, on their own.

* * *

For most of the history of this country our motto, implied or spoken, has been Think Big. I have come to believe that a better motto, and an essential one now, is Think Little. That implies the necessary change of thinking and feeling, and suggests the necessary work. Thinking Big has led us to the two biggest and cheapest political dodges of our time: plan-making and law-making. The lotus-eaters of this era are in Washington, D.C., Thinking Big. Somebody comes up with a problem, and somebody in the government comes up with a plan or a law. The result, mostly, has been the persistence of the problem, and the enlargement and enrichment of the government.

But the discipline of thought is not generalization; it is detail, and it is personal behavior. While the government is "studying" and funding and organizing its Big Thought, nothing is being done. But the citizen who is willing to Think Little, and, accepting the discipline of that, to go ahead on his own, is already solving the problem. A man who is trying to live as a neighbor to his neighbors will have a lively and practical understanding of the work of peace and brotherhood, and let there be no mistake about it—he is *doing* that work. A couple who make a good marriage, and raise healthy, morally competent children, are serving the world's future more directly and surely than any political leader, though they never utter a public word. A good farmer who is dealing with the problem of soil erosion on an acre of ground has a sounder grasp of that problem and *cares* more about it and is probably doing more to solve it than any bureaucrat who is talking about it in general. A man who is willing to undertake the discipline and the difficulty of mending his own ways is worth more to the conservation movement than a hundred who are insisting merely that the government and the industries mend *their* ways.

* * *

If you are concerned about the proliferation of trash, then by all means start an organization in your community to do something about it. But before—*and while*—you organize, pick up some cans and bottles yourself. That way, at least, you will assure yourself and others that you mean what you say. If you are concerned about air pollution, help push for government controls, but drive your car less, use less fuel in your home. If you are worried about the damming of wilderness rivers, join the Sierra Club, write to the government, but turn off the lights you're not using, don't install an air conditioner, don't be a sucker for electrical gadgets, don't waste water. In other words, if you are fearful of the destruction of the environment, then learn to quit being an environmental parasite. We all are, in one way or another, and the remedies are not always obvious, though they certainly will always be difficult. They require a new kind of life—harder, more laborious, poorer in luxuries and gadgets, but also, I am certain, richer in meaning and more abundant in real pleasure. To have a healthy environment we will all have to give up things we like; we may even have to give up things we have come to think of as necessities. But to be fearful of the disease and yet unwilling to pay for the cure is not just to be hypocritical; it is to be doomed. If you talk a good line without being changed by what you say, then you are not just hypocritical and doomed; you have become an agent of the disease. Consider, for an example, President Nixon, who advertises his grave concern about the destruction of the environment, and who turns up the air conditioner to make it cool enough to build a fire.

Odd as I am sure it will appear to some, I can think of no better form of personal involvement in the cure of the environment than that of gardening. A person who is growing a garden, if he is growing it organically, is improving a piece of the world. He is producing something to eat, which makes him somewhat independent of the grocery business, but he is also enlarging, for himself, the meaning of food and the pleasure of eating. The food he grows will be fresher, more nutritious, less contaminated by poisons and preservatives and dyes than what he can buy at a store. He is reducing the trash problem; a garden is not a disposable container, and it will digest and reuse its own wastes. If he enjoys working in his garden, then he is less dependent on an automobile or a merchant for his pleasure. He is involving himself directly in the work of feeding people.

If you think I'm wandering off the subject, let me remind you that most of the vegetables necessary for a family of four can be grown on a plot of forty by sixty feet. I think we might see in this an economic potential of considerable importance, since we now appear to be facing the possibility of widespread famine. How much food could be grown in the dooryards of cities and suburbs? How much could be grown along the extravagant rights-of-way of the interstate system? Or how much could be grown, by the intensive practices and economics of the small farm, on so-called marginal lands? Louis Bromfield liked to point out that the people of France survived crisis after crisis because they were a nation of gardeners, who in times of want turned with great skill to their own small plots of ground. And F. H. King, an agriculture professor who traveled extensively in the Orient in 1907, talked to a Chinese farmer who supported a family of twelve, "one donkey, one cow . . . and two pigs on 2.5 acres of cultivated land"—and who did this, moreover, by agricultural methods that were sound enough organically to have maintained his land in prime fertility through several thousand years of such use. These are possibilities that are readily apparent and attractive to minds that are prepared to Think Little. To Big Thinkers—the bureaucrats and businessmen of agriculture—they are quite simply invisible. But intensive, organic agriculture kept the farms of the Orient thriving for thousands of years, whereas extensive—which is to say, exploitive or extractive—agriculture has critically reduced the fertility of American farmlands in a few centuries or even a few decades.

A person who undertakes to grow a garden at home, by practices that will preserve rather than exploit the economy of the soil, has set his mind decisively against what is wrong with us. He is helping himself in a way that dignifies him and that is rich in meaning and pleasure. But he is doing something else that is more important: he is making vital contact with the soil and the weather on which his life depends. He will no longer look upon rain as an impediment of traffic, or upon the sun as a holiday decoration. And his sense of man's dependence on the world will have grown precise enough, one would hope, to be politically clarifying and useful.

* * *

What I am saying is that if we apply our minds directly and competently to the needs of the earth, then we will have begun to make fundamental and necessary changes in our minds. We will begin to understand and to mistrust *and to change* our wasteful economy, which markets not just the produce of the earth, but also the earth's ability to produce. We will see that beauty and utility are alike dependent upon the health of the world. But we will also see through the fads and the fashions of protest. We will see that war and oppression and pollution are not separate issues, but are aspects of the same issue. Amid the outcries for the liberation of this group or that, we will know that no person is free except in the freedom of other persons, and that man's only real freedom is to know and faithfully occupy his place—a much humbler place than we have been taught to think—in the order of creation.

But the change of mind I am talking about involves not just a change of knowledge, but also a change of attitude toward our essential ignorance, a change in our bearing in the face of mystery. The principle of ecology, if we will take it to heart, should keep us aware that our lives depend upon other lives and upon processes and energies in an interlocking system that, though we can destroy it, we can neither fully understand nor fully control. And our great dangerousness is that, locked in our selfish and myopic economics, we have been willing to change or destroy far beyond our power to understand. We are not humble enough or reverent enough.

Some time ago, I heard a representative of a paper company refer to conservation as a "no-return investment." This man's thinking was exclusively oriented to the annual profit of his industry. Circumscribed by the demand that the profit be great, he simply could not be answerable to any other demand—not even to the obvious needs of his own children.

Consider, in contrast, the profound ecological intelligence of Black Elk, "a holy man of the Oglala Sioux," who in telling his story said that it was not his own life that was important to him, but what he had shared with all life: "It is the story of all life that is holy and it is good to tell, and of us two-leggeds sharing in it with the four-leggeds and the wings of the air and all green things. . . ." And of the great vision that came to him when he was a child he said: "I saw that the sacred hoop of my people was one of many hoops that made one circle, wide as daylight and as starlight, and in the center grew one mighty flowering tree to shelter all the children of one mother and father. And I saw that it was holy."

PART III

The Agrarian Basis for an Authentic Culture

The heart of agrarianism shares the ecological insight that people do not exist in isolation as autonomous beings. We live with and from others in extensive webs of interrelatedness. The following essays describe the nature of our bonds with others, arguing that happiness and health follow from the acceptance of responsibility for our interdependence. Central to Berry's argument is an account of the harmful effects of the breakup of communal life, and the benefits of life together. Agrarian practice demonstrates that our responsibilities extend beyond the human to include the land and the various life forms it supports. The idea that human communities can flourish at the expense of the broad natural communities that sustain them is shown to be false, since it rests on a faulty understanding of the requirements and limits of communal life.

The Body and the Earth

On the Cliff

The question of human limits, of the proper definition and place of human beings within the order of Creation, finally rests upon our attitude toward our biological existence, the life of the body in this world. What value and respect do we give to our bodies? What uses do we have for them? What relation do we see, if any, between body and mind, or body and soul? What connections or responsibilities do we maintain between our bodies and the earth? These are religious questions, obviously, for our bodies are part of the Creation, and they involve us in all the issues of mystery. But the questions are also agricultural, for no matter how urban our life, our bodies live by farming; we come from the earth and return to it, and so we live in agriculture as we live in flesh. While we live our bodies are moving particles of the earth, joined inextricably both to the soil and to the bodies of other living creatures. It is hardly surprising, then, that there should be some profound resemblances between our treatment of our bodies and our treatment of the earth.

That humans are small within the Creation is an ancient perception, represented often enough in art that it must be supposed to have an elemental importance. On one of the painted walls of the Lascaux cave (20,000–15,000 B.C.), surrounded by the exquisitely shaped, shaded, and colored bodies of animals, there is the childish stick figure of a man, a huntsman who, having cast his spear into the guts of a bison, is now weaponless and vulnerable, poignantly frail, exposed, and incomplete. The message seems essentially that of the voice out of the whirlwind in the Book of Job: the Creation is bounteous and mysterious, and humanity is only a part of it—not its equal, much less its master.

Old Chinese landscape paintings reveal, among towering mountains, the frail outline of a roof or a tiny human figure passing along a road on foot or horseback. These landscapes are almost always populated. There is no implication of a dehumanized interest in nature "for its own sake." What is represented is a world in which humans belong, but which does not belong to humans in any tidy economic sense; the Creation provides a place for humans, but it is greater than humanity and within it even great men are small. Such humility is the consequence of an accurate insight, ecological in its bearing, not a pious deference to "spiritual" value.

Closer to us is a passage from the fourth act of *King Lear*, describing the outlook from one of the Dover cliffs:

> The crows and choughs that wing the midway air
> Show scarce so gross as beetles. Halfway down
> Hangs one that gathers samphire, dreadful trade!
> Methinks he seems no bigger than his head.
> The fishermen that walk upon the beach
> Appear like mice, and yond tall anchoring bark
> Diminished to her cock—her cock, a buoy
> Almost too small for sight.

And this is no mere description of a scenic "view." It is part of a play-within-a-play, a sort of ritual of healing. In it Shakespeare is concerned with the curative power of the perception we are dealing with: by understanding accurately his proper place in Creation, a man may be made whole.

In the lines quoted, Edgar, disguised as a lunatic, a Bedlamite, is speaking to his father, the Earl of Gloucester. Gloucester, having been blinded by the treachery of his false son, Edmund, has despaired and has asked the supposed madman to lead him to the cliff's edge, where he intends to destroy himself.

But Edgar's description is from memory; the two are not standing on any such dizzy verge. What we are witnessing is the working out of Edgar's strategy to save his father from false feeling—both the pride, the smug credulity, that led to his suffering and the despair that is its result. These emotions are perceived as madness; Gloucester's blindness is literally the result of the moral blindness of his pride, and it is symbolic of the spiritual blindness of his despair.

Thinking himself on the edge of a cliff, he renounces this world and throws himself down. Though he falls only to the level of his own feet, he is momentarily stunned. Edgar remains with him, but now represents himself as an innocent bystander at the foot of what Gloucester will continue to think is a tall cliff. As the old man recovers his senses, Edgar persuades him that the madman who led him to the cliff's edge was in reality a "fiend." And Gloucester repents his self-destructiveness, which he now recognizes as another kind of pride; a human has no right to destroy what he did not create:

> You ever-gentle gods, take my breath from me.
> Let not my worser spirit tempt me again
> To die before you please.

What Gloucester has passed through, then, is a rite of death and rebirth. In his new awakening he is finally able to recognize his true son. He escapes the unhuman conditions of godly pride and fiendish despair and dies "smilingly" in the truly human estate "'Twixt two extremes of passion, joy and grief . . ."

Until modern times, we focused a great deal of the best of our thought upon such rituals of return to the human condition. Seeking enlightenment or the Promised Land or the way home, a man would go or be forced to go into the wilderness, measure himself against the Creation, recognize finally his true place within it, and thus be saved both from pride and from despair. Seeing himself as a tiny member of a world he cannot comprehend or master or in any final sense possess, he cannot possibly think of himself as a god. And by the same token, since he shares in, depends upon, and is graced by all of which he is a part, neither can be become a fiend; he cannot descend into the final despair of destructiveness. Returning from the wilderness, he becomes a restorer of order, a preserver. He sees the truth, recognizes his true heir, honors his forebears and his heritage, and gives his blessing to his successors. He embodies the passing of human time, living and dying within the human limits of grief and joy.

On the Tower

Apparently with the rise of industry, we began to romanticize the wilderness —which is to say we began to institutionalize it within the concept of the "scenic." Because of railroads and improved highways, the wilderness was no longer an arduous passage for the traveler, but something to be looked at as grand or beautiful from the high vantages of the roadside. We became viewers of "views." And because we no longer traveled in the wilderness as a matter of course, we forgot that wilderness still circumscribed civilization and persisted in domesticity. We forgot, indeed, that the civilized and the domestic continued to *depend* upon wilderness—that is, upon natural forces within the climate and within the soil that have never in any meaningful sense been controlled or conquered. Modern civilization has been built largely in this forgetfulness.

And as we transformed the wilderness into scenery, we began to feel in the presence of "nature" an awe that was increasingly statistical. We would not become appreciators of the Creation until we had taken its measure. Once we had climbed or driven to the mountaintop, we were awed by the view, but it was an awe that we felt compelled to validate or prove by the knowledge of how high we stood and how far we saw. We are invited to "see seven states from atop Lookout Mountain," as if our political boundaries had been drawn in red on the third morning of Creation.

We became less and less capable of sensing ourselves as small within Creation, partly because we thought we could comprehend it statistically, but also because we were becoming creators, ourselves, of a mechanical creation by which we felt ourselves greatly magnified. We built bridges that stood imposingly in titanic settings, towers that stood around us like geologic presences, single machines that could do the work of hundreds of people. Why, after all, should one get excited about a mountain when one can see almost as far from the top of a building, much farther from an airplane, farther still from a space capsule? We have learned to be fascinated by the statistics of magnitude and power. There is apparently no limit in sight, no end, and so it is no wonder that our minds, dizzy with numbers, take refuge in a yearning for infinitudes of energy and materials.

And yet these works that so magnify us also dwarf us, reduce us to insignificance. They magnify us because we are capable of them. They diminish us because, say what we will, once we build beyond a human scale, once we conceive ourselves as Titans or as gods, we are lost in magnitude; we cannot control or limit what we do. The statistics of magnitude call out like Sirens to the statistics of destruction. If we have built towering cities, we have raised even higher the cloud of megadeath. If people are as grass before God, they are as nothing before their machines.

If we are fascinated by the statistics of magnitude, we are no less fascinated by the statistics of our insignificance. We never tire of repeating the commonizing figures of population and population growth. We are entranced to think of ourselves as specks on the pages of our own overwhelming history. I remember that my high-school biology text dealt with the human body by listing its constituent elements, measuring their quantities, and giving their monetary worth — at that time a little less than a dollar. That was a bit of the typical fodder of the modern mind, at once sensational and belittling — no accidental product of the age of Dachau and Hiroshima.

In our time Shakespeare's cliff has become the tower of a bridge — not the scene of a wakening rite of symbolic death and rebirth, but of the real and final death of suicide. Hart Crane wrote its paradigm, as if against his will, in *The Bridge*:

> Out of some subway scuttle, cell or loft
> A bedlamite speeds to thy parapets,
> Tilting there momentarily, shrill shirt ballooning,
> A jest falls from the speechless caravan.

In Shakespeare, the real Bedlamite or madman is the desperate and suicidal Gloucester. The supposed Bedlamite is in reality his true son, and together they enact an eloquent ritual in which Edgar gives his father a vision of Creation. Gloucester abandons himself to this vision, literally casting himself into it, and is renewed; he finds his life by losing it. Gloucester is saved by a renewal of his sense of the world and of his proper place in it. And this is brought about by an enactment that is communal, both in the sense that he is accompanied in it by his son, who for the time being has assumed the disguise of a madman but the role of a priest, and in the sense that it is deeply traditional in its symbols and meanings. In Crane, on the other hand, the Bedlamite is alone, surrounded by speechlessness, cut off within the crowd from any saving or renewing vision. The height, which in Shakespeare is the traditional place of vision, has become in Crane a place of blindness; the bridge, which Crane intended as a unifying symbol, has become the symbol of a final estrangement.

Health

After I had begun to think about these things, I received a letter containing an account of a more recent suicide. The following sentences from that letter seem both to corroborate Crane's lines and to clarify them:

"My friend ______ jumped off the Golden Gate Bridge two months ago. . . . She had been terribly depressed for years. There was no help for her. None that she could find that was sufficient. She was trying to get from one phase of her life to another, and couldn't make it. She had been terribly wounded as a child. . . . Her wound could not be healed. She destroyed herself."

The letter had already asked, "How does a human pass through youth to maturity without 'breaking down'?" And it had answered: "help from tradition, through ceremonies and rituals, rites of passage at the most difficult stages."

My correspondent went on to say: "Healing, it seems to me, is a necessary and useful word when we talk about agriculture." And a few paragraphs later he wrote: "The theme of suicide belongs in a book about agriculture . . ."

I agree. But I am also aware that many people will find it exceedingly strange that these themes should enter so forcibly into this book. It will be thought that I am off the subject. And so I want to take pains to show that I am *on* the subject—and on it, moreover, in the only way most people have of getting on it: by way of the issue of their own health. Indeed, it is when one approaches agriculture from any *other* issue than that of health that one may be said to be off the subject.

The difficulty probably lies in our narrowed understanding of the word *health*. That there is some connection between how we feel and what we eat, between our bodies and the earth, is acknowledged when we say that we must "eat right to keep fit" or that we should eat "a balanced diet." But by health we mean little more than how we feel. We are healthy, we think, if we do not feel any pain or too much pain, and if we are strong enough to do our work. If we become unhealthy, then we go to a doctor who we hope will "cure" us and restore us to health. By health, in other words, we mean merely the absence of disease. Our health professionals are interested almost exclusively in preventing disease (mainly by destroying germs) and in curing disease (mainly by surgery and by destroying germs).

But the concept of health is rooted in the concept of wholeness. To be healthy is to be whole. The word *health* belongs to a family of words, a listing of which will suggest how far the consideration of health must carry us: *heal*, *whole*, *wholesome*, *hale*, *hallow*, *holy*. And so it is possible to give a definition to health that is positive and far more elaborate than that given to it by most medical doctors and the officers of public health.

If the body is healthy, then it is whole. But how can it be whole and yet be dependent, as it obviously is, upon other bodies and upon the earth, upon all the rest of Creation, in fact? It becomes clear that the health or wholeness of the body is a vast subject, and that to preserve it calls for a vast enterprise. Blake said that "Man has no Body distinct from his Soul . . ." and thus acknowledged the convergence of health and holiness. In that, all the convergences and dependences of Creation are surely implied. Our bodies are also not distinct from the bodies of other people, on which they depend in a complexity of ways from biological to spiritual. They are not distinct from the bodies of plants and animals, with which we are involved in the cycles of feeding and in the intricate companionships of ecological systems and of the spirit. They are not distinct from the earth, the sun and moon, and the other heavenly bodies.

It is therefore absurd to approach the subject of health piecemeal with a departmentalized band of specialists. A medical doctor uninterested in nutrition, in agriculture, in the wholesomeness of mind and spirit is as absurd as a farmer who is uninterested in health. Our fragmentation of this subject cannot be our cure, because it is our disease. The body cannot be whole alone. Persons cannot be whole alone. It is wrong to think that bodily health is compatible with spiritual confusion or cultural disorder, or with polluted air and water or impoverished soil. Intellectually, we know that these patterns of interdependence exist; we understand them better now perhaps than we ever have before; yet modern social and cultural patterns contradict them and make it difficult or impossible to honor them in practice.

To try to heal the body alone is to collaborate in the destruction of the body. Healing is impossible in loneliness; it is the opposite of loneliness. Conviviality is healing. To be healed we must come with all the other creatures to the feast of Creation. Together, the above two descriptions of suicides suggest this very powerfully. The setting of both is urban, amid the gigantic works of modern humanity. The fatal sickness is despair, a wound that cannot be healed because it is encapsulated in loneliness, surrounded by speechlessness. Past the scale of the human, our works do not liberate us—they confine us. They cut off access to the wilderness of Creation where we must go to be reborn—to receive the awareness, at once humbling and exhilarating, grievous and joyful, that we are a part of Creation, one with all that we live from and all that, in turn, lives from us. They destroy the communal rites of passage that turn us toward the wilderness and bring us home again.

The Isolation of the Body

Perhaps the fundamental damage of the specialist system—the damage from which all other damages issue—has been the isolation of the body. At some point we began to assume that the life of the body would be the business of grocers and medical doctors, who need take no interest in the spirit, whereas the life of the spirit would be the business of churches, which would have at best only a negative interest in the body. In the same way we began to see nothing wrong with putting the body—most often somebody else's body, but frequently our own—to a task that insulted the mind and demeaned the spirit. And we began to find it easier than ever to prefer our own bodies to the bodies of other creatures and to abuse, exploit, and otherwise hold in contempt those other bodies for the greater good or comfort of our own.

The isolation of the body sets it into direct conflict with everything else in Creation. It gives it a value that is destructive of every other value. That this has happened is paradoxical, for the body was set apart from the soul in order that the soul should triumph over the body. The aim is stated in Shakespeare's Sonnet 146 as plainly as anywhere:

> Poor soul, the center of my sinful earth,
> Lord of these rebel powers that thee array,
> Why dost thou pine within and suffer dearth,
> Painting thy outward walls so costly gay?
> Why so large cost, having so short a lease,
> Dost thou upon thy fading mansion spend?
> Shall worms, inheritors of this excess,
> Eat up thy charge? Is this thy body's end?
> Then, soul, live thou upon thy servant's loss,
> And let that pine to aggravate thy store;
> Buy terms divine in selling hours of dross;
> Within be fed, without be rich no more.
> So shalt thou feed on death, that feeds on men,
> And death once dead, there's no more dying then.

The soul is thus set against the body, to thrive at the body's expense. And so a spiritual economy is devised within which the only law is competition. If the soul is to live in this world only by denying the body, then its relation to worldly life becomes extremely simple and superficial. Too simple and superficial, in fact, to cope in any meaningful or useful way with the world. Spiritual value ceases to have any worldly purpose or force. To fail to employ the body in this world at once for its own good and the good of the soul is to issue an invitation to disorder of the most serious kind.

What was not foreseen in this simple-minded economics of religion was that it is not possible to devalue the body and value the soul. The body, cast loose from the soul, is on its own. Devalued and cast out of the temple, the body does not skulk off like a sick dog to die in the bushes. It sets up a counterpart economy of its own, based also on the law of competition, in which it devalues and exploits the spirit. These two economies maintain themselves at each other's expense, living upon each other's loss, collaborating without cease in mutual futility and absurdity.

You cannot devalue the body and value the soul—or value anything else. The prototypical act issuing from this division was to make a person a slave and then instruct him in religion—a "charity" more damaging to the master than to the slave. Contempt for the body is invariably manifested in contempt for other bodies—the bodies of slaves, laborers, women, animals, plants, the earth itself. Relationships with all other creatures become competitive and exploitive rather than collaborative and convivial. The world is seen and dealt with, not as an ecological community, but as a stock exchange, the ethics of which are based on the tragically misnamed "law of the jungle." This "jungle" law is a basic fallacy of modern culture. The body is degraded and saddened by being set in conflict against the Creation itself, of which all bodies are members, therefore members of each other. The body is thus sent to war against itself.

Divided, set against each other, body and soul drive each other to extremes of misapprehension and folly. Nothing could be more absurd than to despise the body and yet yearn for its resurrection. In reaction to this supposedly religious attitude, we get, not reverence or respect for the body, but another kind of contempt: the desire to comfort and indulge the body with equal disregard for its health. The "dialogue of body and soul" in our time is being carried on between those who despise the body for the sake of its resurrection and those, diseased by bodily extravagance and lack of exercise, who nevertheless desire longevity above all things. These think that they oppose each other, and yet they could not exist apart. They are locked in a conflict that is really their collaboration in the destruction of soul and body both.

What this conflict has done, among other things, is to make it extremely difficult to set a proper value on the life of the body in this world—to believe that it is good, howbeit short and imperfect. Until we are able to say this and know what we mean by it, we will not be able to live our lives in the human estate of grief and joy, but repeatedly will be cast outside in violent swings between pride and despair. Desires that cannot be fulfilled in health will keep us hopelessly restless and unsatisfied.

Competition

By dividing body and soul, we divide both from all else. We thus condemn ourselves to a loneliness for which the only compensation is violence—against other creatures, against the earth, against ourselves. For no matter the distinctions we draw between body and soul, body and earth, ourselves and others—the connections, the dependences, the identities remain. And so we fail to contain or control our violence. It gets loose. Though there are categories of violence, or so we think, there are no categories of victims. Violence against one is ultimately violence against all. The willingness to abuse other bodies is the willingness to abuse one's own. To damage the earth is to damage your children. To despise the ground is to despise its fruit; to despise the fruit is to despise its eaters. The wholeness of health is broken by despite.

If competition is the correct relation of creatures to one another and to the earth, then we must ask why exploitation is not more successful than it is. Why, having lived so long at the expense of other creatures and the earth, are we not healthier and happier than we are? Why does modern society exist under constant threat of the same suffering, deprivation, spite, contempt, and obliteration that it has imposed on other people and other creatures? Why do the health of the body and the health of the earth decline together? And why, in consideration of this decline of our worldly flesh and household, our "sinful earth," are we not healthier in spirit?

It is not necessary to have recourse to statistics to see that the human estate is declining with the estate of nature, and that the corruption of the body is the corruption of the soul. I know that the country is full of "leaders" and experts of various sorts who are using statistics to prove the opposite: that we have more cars, more superhighways, more TV sets, motorboats, prepared foods, etc., than any people ever had before—and are therefore better off than any people ever were before. I can see the burgeoning of this "consumer economy" and can appreciate some of its attractions and comforts. But that economy has an inside and an outside; from the outside there are other things to be seen.

I am writing this in the north-central part of Kentucky on a morning near the end of June. We have had rain for two days, hard rain during the last several hours. From where I sit I can see the Kentucky River swiftening and rising, the water already yellow with mud. I know that inside this city-oriented consumer economy there are many people who will never see this muddy rise and many who will see it without knowing what it means. I know also that there are many who will see it, and know what it means, and not care. If it lasts until the weekend there will be people who will find it as good as clear water for motorboating and waterskiing.

In the past several days I have seen some of the worst-eroded cornfields that I have seen in this country in my life. This erosion is occurring on the cash-rented farms of farmers' widows and city farmers, absentee owners, the doctors and businessmen who buy a farm for the tax breaks or to have "a quiet place in the country" for the weekends. It is the direct result of economic and agricultural policy; it might be said to *be* an economic and agricultural policy. The signs of the "agridollar," big-business fantasy of the Butz mentality are all present: the absenteeism, the temporary and shallow interest of the land-renter, the row-cropping of slopes, the lack of rotation, the plowed-out waterways, the rows running up and down the hills. Looked at from the field's edge, this is ruin, criminal folly, moral idiocy. Looked at from Washington, D.C., from inside the "economy," it is called "free enterprise" and "full production."

And around me here, as everywhere else I have been in this country—in Nebraska, Iowa, Indiana, New York, New England, Tennessee—the farmland is in general decline: fields and whole farms abandoned, given up with their scars unmended, washing away under the weeds and bushes; fine land put to row crops year after year, without rest or rotation; buildings and fences going down; good houses standing empty, unpainted, their windows broken.

And it is clear to anyone who looks carefully at any crowd that we are wasting our bodies exactly as we are wasting our land. Our bodies are fat, weak, joyless, sickly, ugly, the virtual prey of the manufacturers of medicine and cosmetics. Our bodies have become marginal; they are growing useless like our "marginal" land because we have less and less use for them. After the games and idle flourishes of modern youth, we use them only as shipping cartons to transport our brains and our few employable muscles back and forth to work.

As for our spirits, they seem more and more to comfort themselves by buying things. No longer in need of the exalted drama of grief and joy, they feed now on little shocks of greed, scandal, and violence. For many of the churchly, the life of the spirit is reduced to a dull preoccupation with getting to Heaven. At best, the world is no more than an embarrassment and a trial to the spirit, which is otherwise radically separated from it. The true lover of God must not be burdened with any care or respect for His works. While the body goes about its business of destroying the earth, the soul is supposed to lie back and wait for Sunday, keeping itself free of earthly contaminants. While the body exploits other bodies, the soul stands aloof, free from sin, crying to the gawking bystanders: "I am not enjoying it!" As far as this sort of "religion" is concerned, the body is no more than the lusterless container of the soul, a mere "package," that will nevertheless light up in eternity, forever cool and shiny as a neon cross. This separation of the soul from the body and from the world is no disease of the fringe, no aberration, but a fracture that runs through the mentality of institutional religion like a geologic fault. And this rift in the mentality of religion continues to characterize the modern mind, no matter how secular or worldly it becomes.

But I have not stated my point exactly enough. This rift is not *like* a geologic fault; it *is* a geologic fault. It is a flaw in the mind that runs inevitably into the earth. Thought affects or afflicts substance neither by intention nor by accident, but because, occurring in the Creation that is unified and whole, it must; there is no help for it.

The soul, in its loneliness, hopes only for "salvation." And yet what is the burden of the Bible if not a sense of the mutuality of influence, rising out of an essential unity, among soul and body and community and world? These are all the works of God, and it is therefore the work of virtue to make or restore harmony among them. The world is certainly thought of as a place of spiritual trial, but it is also the confluence of soul and body, word and flesh, where thoughts must become deeds, where goodness is to be enacted. This is the great meeting place, the narrow passage where spirit and flesh, word and world, pass into each other. The Bible's aim, as I read it, is not the freeing of the spirit from the world. It is the handbook of their interaction. It says that they cannot be divided; that their mutuality, their unity, is inescapable; that they are not reconciled in division, but in harmony. What else can be meant by the resurrection of the body? The body should be "filled with light," perfected in understanding. And so everywhere there is the sense of consequence, fear and desire, grief and joy. What is desirable is repeatedly defined in the tensions of the sense of consequence. False prophets are to be known "by their fruits." We are to treat others as we would be treated; thought is thus barred from any easy escape into aspiration or ideal, is turned around and forced into action. The following verses from Proverbs are not very likely the original work of a philosopher-king; they are overheard from generations of agrarian grandparents whose experience taught them that spiritual qualities become earthly events:

> I went by the field of the slothful, and by the vineyard of the man void of understanding;
>
> And, lo, it was all grown over with thorns, and nettles had covered the face thereof, and the stone wall thereof was broken down.
>
> Then I saw, and considered it well. I looked upon it, and received instruction.
>
> Yet a little sleep, a little slumber, a little folding of the hands to sleep:
>
> So shall thy poverty come as one that traveleth; and thy want as an armed man.

Connections

I do not want to speak of unity misleadingly or too simply. Obvious distinctions can be made between body and soul, one body and other bodies, body and world, etc. But these things that appear to be distinct are nevertheless caught in a network of mutual dependence and influence that is the substantiation of their unity. Body, soul (or mind or spirit), community, and world are all susceptible to each other's influence, and they are all conductors of each other's influence. The body is damaged by the bewilderment of the spirit, and it conducts the influence of that bewilderment into the earth, the earth conducts it into the community, and so on. If a farmer fails to understand what health is, his farm becomes unhealthy; it produces unhealthy food, which damages the health of the community. But this is a network, a spherical network, by which each part is connected to every other part. The farmer is a part of the community, and so it is as impossible to say exactly where the trouble began as to say where it will end. The influences go backward and forward, up and down, round and round, compounding and branching as they go. All that is certain is that an error introduced anywhere in the network ramifies beyond the scope of prediction; consequences occur all over the place, and each consequence breeds further consequences. But it seems unlikely that an error can ramify endlessly. It spreads by way of the connections in the network, but sooner or later it must also begin to break them. We are talking, obviously, about a circulatory system, and a disease of a circulatory system tends first to impair circulation and then to stop it altogether.

Healing, on the other hand, complicates the system by opening and restoring connections among the various parts—in this way restoring the ultimate simplicity of their union. When all the parts of the body are working together, are under each other's influence, we say that it is whole; it is healthy. The same is true of the world, of which our bodies are parts. The parts are healthy insofar as they are joined harmoniously to the whole.

What the specialization of our age suggests, in one example after another, is not only that fragmentation is a disease, but that the diseases of the disconnected parts are similar or analogous to one another. Thus they memorialize their lost unity, their relation persisting in their disconnection. Any severance produces two wounds that are, among other things, the record of how the severed parts once fitted together.

The so-called identity crisis, for instance, is a disease that seems to have become prevalent after the disconnection of body and soul and the other piecemealings of the modern period. One's "identity" is apparently the immaterial part of one's being—also known as psyche, soul, spirit, self, mind, etc. The dividing of this principle from the body and from any particular worldly locality would seem reason enough for a crisis. Treatment, it might be thought, would logically consist in the restoration of these connections: the lost identity would find itself by recognizing physical landmarks, by connecting itself responsibly to practical circumstances; it would learn to stay put in the body to which it belongs and in the place to which preference or history or accident has brought it; it would, in short, find itself in finding its work. But "finding yourself," the pseudo-ritual by which the identity crisis is supposed to be resolved, makes use of no such immediate references. Leaving aside the obvious, and ancient, realities of doubt and self-doubt, as well as the authentic madness that is often the result of cultural disintegration, it seems likely that the identity crisis is a conventional illusion, one of the genres of self-indulgence. It can be an excuse for irresponsibility or a fashionable mode of self-dramatization. It is the easiest form of self-flattery—a way to construe procrastination as a virtue—based on the romantic assumption that "who I really am" is better in some fundamental way than the available evidence would suggest.

The fashionable cure for this condition, if I understand the lore of it correctly, has nothing to do with the assumption of responsibilities or the renewal of connections. The cure is "autonomy," another illusory condition, suggesting that the self can be self-determining and independent without regard for any determining circumstance or any of the obvious dependences. This seems little more than a jargon term for indifference to the opinions and feelings of other people. There is, in practice, no such thing as autonomy. Practically, there is only a distinction between responsible and irresponsible dependence. Inevitably failing this impossible standard of autonomy, the modern self-seeker becomes a tourist of cures, submitting his quest to the guidance of one guru after another. The "cure" thus preserves the disease.

It is not surprising that this strange disease of the spirit—the self's loss of self—should have its counterpart in an anguish of the body. One of the commonplaces of modern experience is dissatisfaction with the body—not as one has allowed it to become, but as it naturally is. The hardship is perhaps greater here because the body, unlike the self, is substantial and cannot be supposed to be inherently better than it was born to be. It can only be thought inherently worse than it *ought* to be. For the appropriate standard for the body—that is, health—has been replaced, not even by another standard, but by very exclusive physical *models*. The concept of "model" here conforms very closely to the model of the scientists and planners: it is an exclusive, narrowly defined ideal that affects destructively whatever it does not include.

Thus our young people are offered the ideal of health only by what they know to be lip service. What they are made to feel forcibly, and to measure themselves by, is the exclusive desirability of a certain physical model. Girls are taught to want to be leggy, slender, large-breasted, curly-haired, unimposingly beautiful. Boys are instructed to be "athletic" in build, tall but not too tall, broad-shouldered, deep-chested, narrow-hipped, square-jawed, straight-nosed, not bald, unimposingly handsome. Both sexes should look what passes for "sexy" in a bathing suit. Neither, above all, should look old.

Though many people, in health, are beautiful, very few resemble these models. The result is widespread suffering that does immeasurable damage both to individual persons and to the society as a whole. The result is another absurd pseudo-ritual, "accepting one's body," which may take years or may be the distraction of a lifetime. Woe to the man who is short or skinny or bald. Woe to the man with a big nose. Woe, above all, to the woman with small breasts or a muscular body or strong features; Homer and Solomon might have thought her beautiful, but she will see her own beauty only by a difficult rebellion. And like the crisis of identity, this crisis of the body brings a helpless dependence on cures. One spends one's life dressing and "making up" to compensate for one's supposed deficiencies. Again, the cure preserves the disease. And the putative healer is the guru of style and beauty aid. The sufferer is by definition a customer.

Sexual Division

To divide body and soul, or body and mind, is to inaugurate an expanding series of divisions—not, however, an *infinitely* expanding series, because it is apparently the nature of division sooner or later to destroy what is divided; the principle of durability is unity. The divisions issuing from the division of body and soul are first sexual and then ecological. Many other divisions branch out from those, but those are the most important because they have to do with the fundamental relationships—with each other and with the earth—that we all have in common.

To think of the body as separate from the soul or as soulless, either to subvert its appetites or to "free" them, is to make an object of it. As a thing, the body is denied any dimension or rightful presence or claim in the mind. The concerns of the body—all that is comprehended in the term *nurture*—are thus degraded, denied any respected place among the "higher things" and even among the more exigent practicalities.

The first sexual division comes about when nurture is made the exclusive concern of women. This cannot happen until a society becomes industrial; in hunting and gathering and in agricultural societies, men are of necessity also involved in nurture. In those societies there usually have been differences between the work of men and that of women. But the necessity here is to distinguish between sexual difference and sexual division.

In an industrial society, following the division of body and soul, we have at the "upper" or professional level a division between "culture" (in the specialized sense of religion, philosophy, art, the humanities, etc.) and "practicality," and both of these become increasingly abstract. Thinkers do not act. And the "practical" men do not work with their hands, but manipulate the abstract quantities and values that come from the work of "workers." Workers are simplified or specialized into machine parts to do the wage-work of the body, which they were initially permitted to think of as "manly" because for the most part women did not do it.

Women traditionally have performed the most confining—though not necessarily the least dignified—tasks of nurture: housekeeping, the care of young children, food preparation. In the urban-industrial situation the confinement of these traditional tasks divided women more and more from the "important" activities of the new economy. Furthermore, in this situation the traditional nurturing role of men—that of provisioning the household, which in an agricultural society had become as constant and as complex as the women's role—became completely abstract; the man's duty to the household came to be simply to provide money. The only remaining *task* of provisioning—purchasing food—was turned over to women. This determination that nurturing should become *exclusively* a concern of women served to signify to both sexes that neither nurture nor womanhood was very important.

But the assignment to women of a kind of work that was thought both onerous and trivial was only the beginning of their exploitation. As the persons exclusively in charge of the tasks of nurture, women often came into sole charge of the household budget; they became family purchasing agents. The time of the household barterer was past. Kitchens were now run on a cash economy. Women had become customers, a fact not long wasted on the salesmen, who saw that in these women they had customers of a new and most promising kind. The modern housewife was isolated from her husband, from her school-age children, and from other women. She was saddled with work from which much of the skill, hence much of the dignity, had been withdrawn, and which she herself was less and less able to consider important. She did not know what her husband did at work, or after work, and she knew that her life was passing in his regardlessness and in his absence. Such a woman was ripe for a sales talk: this was the great commercial insight of modern times. Such a woman must be told—or subtly made to understand—that she must not be a drudge, that she must not let her work affect her looks, that she must not become "unattractive," that she must always be fresh, cheerful, young, shapely, and pretty. All her sexual and mortal fears would thus be given voice, and she would be made to reach for money. What was implied was always the question that a certain bank finally asked outright in a billboard advertisement: "Is your husband losing interest?"

Motivated no longer by practical needs, but by loneliness and fear, women began to identify themselves by what they bought rather than by what they did. They bought labor-saving devices that worked, as most modern machines have tended to work, to devalue or replace the skills of those who used them. They bought manufactured foods, which did likewise. They bought any product that offered to lighten the burdens of housework, to be "kind to hands," or to endear one to one's husband. And they furnished their houses, as they made up their faces and selected their clothes, neither by custom nor invention, but by the suggestion of articles and advertisements in "women's magazines." Thus housewifery, once a complex discipline acknowledged to be one of the bases of culture and economy, was reduced to the exercise of purchasing power.* The housewife's only remaining productive capacity was that of reproduction. But even as a mother she remained a consumer,† subjecting herself to an all-presuming doctor and again to written instructions calculated to result in the purchase of merchandise. Breast-feeding of babies became unfashionable, one suspects, because it was the last form of home production; no way could be found to persuade a woman to purchase her own milk. All these "improvements" involved a radical simplification of mind that was bound to have complicated, and ironic, results. As housekeeping became simpler and easier, it also became more boring. A woman's work became less accomplished and less satisfying. It became easier for her to believe that what she did was not important. And this heightened her anxiety and made her even more avid and even less discriminating as a consumer. The cure not only preserved the disease, it compounded it.

There was, of course, a complementary development in the minds of men, but there is less to say about it. The man's mind was not simplified by a degenerative process, but by a kind of coup: as soon as he separated working and living and began to work away from home, the practical considerations of the household were excerpted from his mind all at once.

In modern marriage, then, what was once a difference of work became a division of work. And in this division the household was destroyed as a practical bond between husband and wife. It was no longer a condition, but only a place. It was no longer a circumstance that required, dignified, and rewarded the enactment of mutual dependence, but the site of mutual estrangement.

*She did continue to do "housework," of course. But we must ask what this had come to mean. The industrial economy had changed the criterion of housekeeping from thrift to convenience. Thrift was a complex standard, requiring skill, intelligence, and moral character, and private thrift was rightly considered a public value. Once thrift was destroyed as a value, housekeeping became simply a corrupt function of a corrupt economy: its public "value" lay in the wearing out or using up of commodities.

†This was written before the era of commercial child production. A woman may now "work" as a brood animal and sell her offspring. This cottage industry is a grand advance for liberty, since it frees some women from impecuniousness and others from sterility—the rule being that one has the right to be freed from any objectionable condition by any means. It is also the most significant expansion of commerce since the Emancipation Proclamation.

Home became a place for the husband to go when he was not working or amusing himself. It was the place where the wife was held in servitude.

A sexual difference is not a wound, or it need not be; a sexual division is. And it is important to recognize that this division — this destroyed household that now stands between the sexes — is a wound that is suffered inescapably by both men and women. Sometimes it is assumed that the estrangement of women in their circumscribed "women's world" can only be for the benefit of men. But that interpretation seems to be based on the law of competition that is modeled in the exploitive industrial economy. This law holds that for everything that is exploited or oppressed there must be something else that is proportionately improved; thus, men must be as happy as women are unhappy.

There is no doubt that women have been deformed by the degenerate housewifery that is now called their "role" — but not, I think, for any man's benefit. If women are deformed by their role, then, insofar as the roles are divided, men are deformed by theirs. Degenerate housewifery is indivisible from degenerate husbandry. There is no escape. This is the justice that we are learning from the ecologists: you cannot damage what you are dependent upon without damaging yourself. The suffering of women is noticed now, is noticeable now, because it is not given any considerable status or compensation. If we removed the status and compensation from the destructive exploits we classify as "manly," men would be found to be suffering as much as women. They would be found to be suffering for the same reason: they are in exile from the communion of men and women, which is their deepest connection with the communion of all creatures.

For example: a man who is in the traditional sense a *good* farmer is husbandman and husband, the begetter and conserver of the earth's bounty, but he is also midwife and motherer. He is a nurturer of life. His work is domestic; he is bound to the household. But let "progress" take such a man and transform him into a technologist of production (that is, sever his bonds to the household, make useless or pointless or "uneconomical" his impulse to conserve and to nurture), and it will have made of him a creature as deformed, and as pained, as it has notoriously made of his wife.

The Dismemberment of the Household

We are familiar with the concept of the disintegral life of our time as a dismembered cathedral, the various concerns of culture no longer existing in reference to each other or within the discipline of any understanding of their unity. It may also be conceived, and its strains more immediately felt, as a dismembered household. Without the household—not just as a unifying ideal, but as a practical circumstance of mutual dependence and obligation, requiring skill, moral discipline, and work—husband and wife find it less and less possible to imagine and enact their marriage. Without much in particular that they can *do* for each other, they have a scarcity of practical reasons to be together. They may "like each other's company," but that is a reason for friendship, not for marriage. Aside from affection for any children they may have and their abstract legal and economic obligations to each other, their union has to be empowered by sexual energy alone.

Perhaps the most dangerous, certainly the most immediately painful, consequence of the disintegration of the household is this isolation of sexuality. The division of sexual energy from the functions of household and community that it ought both to empower and to grace is analogous to that other modern division between hunger and the earth. When it is no longer allied by proximity and analogy to the nurturing disciplines that bound the household to the cycles of fertility and the seasons, life and death, then sexual love loses its symbolic or ritualistic force, its deepest solemnity and its highest joy. It loses its sense of consequence and responsibility. It becomes "autonomous," to be valued only for its own sake, therefore frivolous, therefore destructive—even of itself. Those who speak of sex as "recreation," thinking to claim for it "a new place," only acknowledge its displacement from Creation.

The isolation of sexuality makes it subject to two influences that dangerously oversimplify it: the lore of sexual romance and capitalist economics. By "sexual romance" I mean the sentimentalization of sexual love that for generations has been the work of popular songs and stories. By means of them, young people have been taught a series of extremely dangerous falsehoods:

1. That people in love ought to conform to the fashionable models of physical beauty, and that to be unbeautiful by these standards is to be unlovable.
2. That people in love are, or ought to be, young—even though love is said to last "forever."
3. That marriage is a solution—whereas the most misleading thing a love story can do is to end "happily" with a marriage, not because there is no such thing as a happy marriage, but because marriage cannot be happy except by being *made* happy.
4. That love, alone, regardless of circumstances, can make harmony and resolve serious differences.

5. That "love will find a way" and so finally triumph over any kind of practical difficulty.
6. That the "right" partners are "made for each other," or that "marriages are made in Heaven."
7. That lovers are "each other's all" or "all the world to each other."
8. That monogamous marriage is therefore logical and natural, and "forsaking all others" involves no difficulty.

Believing these things, a young couple could not be more cruelly exposed to the abrasions of experience—or better prepared to experience marriage as another of those grim and ironic modern competitions in which the victory of one is the defeat of both.

As experience frets away gullibility, the exclusiveness of the sentimental ideal gives way to the possessiveness of sexual capitalism. Failing, as they cannot help but fail, to be each other's all, the husband and wife become each other's only. The sacrament of sexual union, which in the time of the household was a communion of workmates, and afterward tried to be a lovers' paradise, has now become a kind of marketplace in which husband and wife represent each other as sexual property. Competitiveness and jealousy, imperfectly sweetened and disguised by the illusions of courtship, now become governing principles, and they work to isolate the couple inside their marriage. Marriage becomes a capsule of sexual fate. The man must look on other men, and the woman on other women, as threats. This seems to have become particularly damaging to women; because of the progressive degeneration and isolation of their "role," their worldly stock-in-trade has increasingly had to be "their" men. In the isolation of the resulting sexual "privacy," the disintegration of the community begins. The energy that is the most convivial and unifying loses its communal forms and becomes divisive. This dispersal was nowhere more poignantly exemplified than in the replacement of the old ring dances, in which all couples danced together, by the so-called ballroom dancing, in which each couple dances alone. A significant part of the etiquette of ballroom dancing is, or was, that the exchange of partners was accomplished by a "trade." It is no accident that this capitalization of love and marriage was followed by a divorce epidemic—and by fashions of dancing in which each one of the dancers moves alone.

The disintegration of marriage, which completes the disintegration of community, came about because the encapsulation of sexuality, meant to preserve marriage from competition, inevitably *enclosed* competition. The principle that fenced everyone else out fenced the couple in; it became a sexual cul-desac. The model of economic competition proved as false to marriage as to farming. As with other capsules, the narrowness of the selective principle proved destructive of what it excluded, and what it excluded was essential to the life of what it enclosed: the nature of sexuality itself. Sexual romance cannot bear to acknowledge the generality of instinct, whereas sexual capitalism cannot acknowledge its particularity. But sexuality appears to be *both* general and particular. One cannot love a particular woman, for instance, unless one loves womankind—if not all women, at least other women. The capsule of sexual romance leaves out this generality, this generosity of instinct; it excludes Aphrodite and Dionysus. And it fails for that reason. Though sexual love can endure between the same two people for a long time, it cannot do so on the basis of this pretense of the exclusiveness of affection. The sexual capitalist—that is, the disillusioned sexual romantic—in reaction to disillusion makes the opposite oversimplification; one acknowledges one's spouse as one of a general, necessarily troublesome kind or category.

Both these attitudes look on sexual love as ownership. The sexual romantic croons, "You be-long to me." The sexual capitalist believes the same thing but has stopped crooning. Each holds that a person's sexual property shall be sufficient unto him or unto her, and that the morality of that sufficiency is to be forever on guard against expropriation. Within the capsule of marriage, as in that of economics, one intends to exploit one's property and to protect it. Once the idea of property becomes abstract or economic, both these motives begin to rule over it. They are, of course, contradictory; all that one can really protect is one's "right" or intention to exploit. The proprieties and privacies used to encapsulate marriage may have come from the tacit recognition that exploitive sex, like exploitive economics, is a very dirty business. One makes a secret of the sexuality of one's marriage for the same reason that one posts "Keep Out/Private Property" on one's strip mine. The tragedy, more often felt than acknowledged, is that what is exploited becomes undesirable.

The protective capsule becomes a prison. It becomes a household of the living dead, each body a piece of incriminating evidence. Or a greenhouse excluding the neighbors and the weather for the sake of some alien and unnatural growth. The marriage shrinks to a dull vigil of duty and legality. Husband and wife become competitors necessarily, for their only freedom is to exploit each other or to escape.

It is possible to imagine a more generous enclosure—a household welcoming to neighbors and friends; a garden open to the weather, between the woods and the road. It is possible to imagine a marriage bond that would bind a woman and a man not only to each other, but to the community of marriage, the amorous communion at which all couples sit: the sexual feast and celebration that joins them to all living things and to the fertility of the earth, and the sexual responsibility that joins them to the human past and the human future. It is possible to imagine marriage as a grievous, joyous human bond, endlessly renewable and renewing, again and again rejoining memory and passion and hope.

Fidelity

But it is extremely difficult, now, to imagine marriage in terms of such dignity and generosity, and this difficulty is explained by the failure of these possessive and competitive forms of sexual love that have been in use for so long. This failure raises unavoidably the issue of fidelity: What is it, and what does it mean—in marriage, and also, since marriage is a fundamental relationship and metaphor, in other relationships?

No one can be glad to have this issue so starkly raised, for any consideration of it now must necessarily involve one's own bewilderment. We are apparently near the end of a degenerative phase of an evolutionary process—a long way from any large-scale regeneration. For that reason it is necessary to be hesitant and cautious, respectful of the complexity and importance of the problem. Marriage is not going to change because somebody thinks about it and recommends an "answer"; it can change only as its necessities are felt and as its circumstances change.

The idea of fidelity is perverted beyond redemption by understanding it as a grim, literal duty enforced only by willpower. This is the "religious" insanity of making a victim of the body as a victory of the soul. Self-restraint that is so purely negative is self-hatred. And one cannot be good, anyhow, just by not being bad. To be faithful merely out of duty is to be blinded to the possibility of a better faithfulness for better reasons.

It is reasonable to suppose, if fidelity is a virtue, that it is a virtue with a purpose. A purposeless virtue is a contradiction in terms. Virtue, like harmony, cannot exist alone; a virtue must lead to harmony between one creature and another. To be good for nothing is just that. If a virtue has been thought a virtue long enough, it must be assumed to have practical justification—though the very longevity that proves its practicality may obscure it. That seems to be what happened with the idea of fidelity. We heard the words "forsaking all others" repeated over and over again for so long that we lost the sense of their practical justification. They assumed the force of superstition: people came to be faithful in marriage not out of any understanding of the meaning of faith or of marriage, but out of the same fear of obscure retribution that made one careful not to break a mirror or spill the salt. Like other superstitions, this one was weakened by the scientific, positivist intellectuality of modern times and by the popular "sophistication" that came with it. Our age could be characterized as a manifold experiment in faithlessness, and if it has as yet produced no effective understanding of the practicalities of faith, it has certainly produced massive evidence of the damage and disorder of its absence.

It is possible to open this issue of the practicality of fidelity by considering that the modern age was made possible by the freeing, and concurrently by the cheapening, of energy. It can be said, of course, that the modern age was made possible by technologies that *control* energy and thus make it usable at an unprecedented rate. But such control is at best extremely limited: the devices by which industrial and military energies are used control them only momentarily; their moment of usefulness sets them loose in the world as social, ecological, and geological *forces*. We can use these energies only as explosives; we can control the rate, intensity, and time of combustion, but our effective control ends with the use of the small amount of the released energy that we are able to harness. Past that, the effects are on their own, to compound themselves as they will. In modern times we have never been able to subject our use of energy to a sense of responsibility anywhere near complex enough to be equal to its effects.

It may be that the principle of sexual fidelity, once it is again fully understood, will provide us with as good an example as we can find of the responsible use of energy. Sexuality is, after all, a form of energy, one of the most powerful. If we see sexuality as energy, then it becomes impossible to see sexual fidelity as merely a "duty," a virtue for the sake of virtue, or a superstition. If we made a superstition of fidelity, and thereby weakened it, by thinking of it as purely a moral or spiritual virtue, then perhaps we can restore its strength by recovering an awareness of its practicality.

At the root of culture must be the realization that uncontrolled energy is disorderly—that in nature all energies move in forms; that, therefore, in a human order energies must be *given* forms. It must have been plain at the beginning, as cultural degeneracy has made it plain again and again, that one can be indiscriminately sexual but not indiscriminately responsible, and that irresponsible sexuality would undermine any possibility of culture since it implies a hierarchy based purely upon brute strength, cunning, regardlessness of value and of consequence. Fidelity can thus be seen as the necessary discipline of sexuality, the practical definition of sexual responsibility, or the definition of the moral limits within which such responsibility can be conceived and enacted. The forsaking of all others is a keeping of faith, not just with the chosen one, but with the ones forsaken. The marriage vow unites not just a woman and a man with each other; it unites each of them with the community in a vow of sexual responsibility toward all others. The whole community is married, realizes its essential unity, in each of its marriages.*

Another use of fidelity is to preserve the possibility of devotion against the distractions of novelty. What marriage offers—and what fidelity is meant to protect—is the possibility of moments when what we have chosen and what we desire are the same. Such a convergence obviously cannot be continuous. No relationship can continue very long at its highest emotional pitch. But fidelity prepares us for the *return* of these moments, which give us the highest joy we can know: that of union, communion, atonement (in the root sense of at-one-ment). The principle is stated in these lines by William Butler Yeats (by "the world" he means the world after the Fall):

> Maybe the bride-bed brings despair,
> For each an imagined image brings
> And finds a real image there;
> Yet the world ends when these two things,
> Though several, are a single light . . .

To forsake all others does not mean—because it *cannot* mean—to ignore or neglect all others, to hide or be hidden from all others, or to desire or love no others. To live in marriage is a responsible way to live in sexuality, as to live in a household is a responsible way to live in the world. One cannot enact or fulfill one's love for womankind or mankind, or even for all the women or men to whom one is attracted. If one is to have the power and delight of one's sexuality, then the generality of instinct must be resolved in a responsible relation-

*Marital fidelity, that is, involves the public or institutional as well as the private aspect of marriage. One is married to marriage as well as to one's spouse. But one is married also to something vital of one's own that does not exist before the marriage: one's given word. It now seems to me that the modern misunderstanding of marriage involves a gross misunderstanding and underestimation of the seriousness of giving one's word, and of the dangers of breaking it once it is given. Adultery and divorce now must be looked upon as instances of that disease of word-breaking, which our age justifies as "realistic" or "practical" or "necessary," but which is tattering the invariably single fabric of speech and trust.

ship to a particular person. Similarly, one cannot live in the world; that is, one cannot become, in the easy, generalizing sense with which the phrase is commonly used, a "world citizen." There can be no such thing as a "global village." No matter how much one may love the world as a whole, one can live fully in it only by living responsibly in some small part of it. Where we live and who we live there with define the terms of our relationship to the world and to humanity. We thus come again to the paradox that one can become whole only by the responsible acceptance of one's partiality.

But to encapsulate these partial relationships is to entrap and condemn them in their partiality; it is to endanger them and to make them dangerous. They are enlivened and given the possibility of renewal by the double sense of particularity and generality: one lives in marriage *and* in sexuality, at home *and* in the world. It is impossible, for instance, to conceive that a man could despise women and yet love his wife, or love his own place in the world and yet deal destructively with other places.

Home Land and House Hold

What I have been trying to do is to define a pattern of disintegration that is at once cultural and agricultural. I have been groping for connections—that I think are indissoluble, though obscured by modern ambitions—between the spirit and the body, the body and other bodies, the body and the earth. If these connections do necessarily exist, as I believe they do, then it is impossible for material order to exist side by side with spiritual disorder, or vice versa, and impossible for one to thrive long at the expense of the other; it is impossible, ultimately, to preserve ourselves apart from our willingness to preserve other creatures, or to respect and care for ourselves except as we respect and care for other creatures; and, most to the point of this book, it is impossible to care for each other more or differently than we care for the earth.

This last statement becomes obvious enough when it is considered that the earth is what we all have in common, that it is what we are made of and what we live from, and that we therefore cannot damage it without damaging those with whom we share it. But I believe it goes farther and deeper than that. There is an uncanny *resemblance* between our behavior toward each other and our behavior toward the earth. Between our relation to our own sexuality and our relation to the reproductivity of the earth, for instance, the resemblance is plain and strong and apparently inescapable. By some connection that we do not recognize, the willingness to exploit one becomes the willingness to exploit the other. The conditions and the means of exploitation are likewise similar.

The modern failure of marriage that has so estranged the sexes from each other seems analogous to the "social mobility" that has estranged us from our land, and the two are historically parallel. It may even be argued that these two estrangements are very close to being one, both of them having been caused by the disintegration of the household, which was the formal bond between marriage and the earth, between human sexuality and its sources in the sexuality of Creation. The importance of this practical bond has not been often or very openly recognized in our tradition; in modern times it has almost disappeared under the burden of adverse fashion and economics. It is necessary to go far back to find it clearly exemplified.

To my mind, one of the best examples that we have is in Homer's *Odyssey*. Nowhere else that I know are the connections between marriage and household and the earth so fully and so carefully understood.

At the opening of the story Odysseus, after a twenty-year absence, is about to begin the last leg of his homeward journey. The sole survivor of all his company of warriors, having lived through terrible trials and losses, Odysseus is now a castaway on the island of the goddess Kalypso. He is Kalypso's lover but also virtually her prisoner. At night he sleeps with Kalypso in her cave; by day he looks across the sea toward Ithaka, his home, and weeps. Homer does not stint either feeling—the delights of Kalypso's cave, where the lovers "revel and rest softly, side by side," or the grief and longing of exile.

But now Zeus commands Kalypso to allow Odysseus to depart; she comes to tell him that he is free to go. And yet it is a tragic choice that she offers him: he must choose between her and Penélopê, his wife. If he chooses Kalypso, he will be immortal, but remain in exile; if he chooses Penélopê, he will return home at last, but will die in his time like other men:

> If you could see it all, before you go—
> all the adversity you face at sea—
> you would stay here, and guard this house, and be
> immortal—though you wanted her forever,
> that bride for whom you pine each day.
> Can I be less desirable than she is?
> Less interesting? Less beautiful? Can mortals
> compare with goddesses in grace and form?

And Odysseus answers:

> My quiet Penélopê — how well I know —
> would seem a shade before your majesty,
> death and old age being unknown to you,
> while she must die. Yet, it is true, each day
> I long for home . . .

This is, in effect, a wedding ritual much like our own, in which Odysseus forsakes all others, in renouncing the immortal womanhood of the goddess, and renews his pledge to the mortal terms of his marriage. But unlike our ritual, this one involves an explicit loyalty to a home. Odysseus's far-wandering through the wilderness of the sea is not merely the return of a husband; it is a journey home. And a great deal of the power as well as the moral complexity of *The Odyssey* rises out of the richness of its sense of home.

By the end of Book XXIII, it is clear that the action of the narrative, Odysseus's journey from the cave of Kalypso to the bed of Penélopê, has revealed a structure that is at once geographical and moral. This structure may be graphed as a series of diminishing circles centered on one of the posts of the marriage bed. Odysseus makes his way from the periphery toward that center.

All around, this structure verges on the sea, which is the wilderness, ruled by the forces of nature and by the gods. In spite of the excellence of his ship and crew and his skill in navigation, a man is alien there. Only when he steps ashore does he enter a human order. From the shoreline of his island of Ithaka, Odysseus makes his way across a succession of boundaries, enclosed and enclosing, with the concentricity of a blossom around its pistil, a human pattern resembling a pattern of nature. He comes to his island, to his own lands, to his town, to his household and house, to his bedroom, to his bed.

As he moves toward this center he moves also through a series of recognitions, tests of identity and devotion. By these, his homecoming becomes at the same time a restoration of order. At first, having been for a while uncertain of his whereabouts, he recognizes his homeland by the conformation of the countryside and by a certain olive tree. He then becomes the guest of his swineherd, Eumaios, and tests his loyalty, though Eumaios will not be permitted to recognize his master until the story approaches its crisis. In the house of Eumaios, Odysseus meets and makes himself known to his son, Telémakhos. As he comes, disguised as a beggar, into his own house, he is recognized by Argus, his old hunting dog. That night, as the guest of Penélopê, who does not yet know who he is, he is recognized by his aged nurse, Eurükleia, who sees a well-remembered scar on his thigh as she is bathing his feet.

He is scorned and abused as a vagabond by the band of suitors who, believing him dead, have been courting his wife, consuming his meat and wine, desecrating his household, and plotting the murder of his son. Penélopê proposes a trial by which the suitors will compete for her: she will become the bride of whichever one can string the bow of her supposedly dead husband and shoot an arrow through the aligned helve-sockets of twelve axe heads. The suitors fail. Odysseus performs the feat easily and is thereby recognized as "the great husband" himself. And then, with the help of the swineherd, the cowherd, and Telémakhos, he proceeds to trap the suitors and slaughter them all without mercy. To so distinguished a commentator as Richmond Lattimore, their punishment "seems excessive." But granting the acceptability of violent means to a warrior such as Odysseus, this outcome seems to me appropriate to the moral terms of the poem. It is made clear that the punishment is not merely the caprice of a human passion: Odysseus enacts the will of the gods; he is the agent of a divine judgment. The suitors' sin is their utter contempt for the domestic order that the poem affirms. They do not respect or honor the meaning of the household, and in *The Odyssey* this meaning is paramount.

It is therefore the recognition of Odysseus by Penélopê that is the most interesting and the most crucial. By the time Odysseus's vengeance and his purification of the house are complete, Penélopê is the only one in the household who has not acknowledged him. It is only reasonable that she should delay this until she is absolutely certain. After all, she has waited twenty years; it is not to be expected that she would be less than cautious now. Her faith has been equal and more than equal to his, and now she proves his equal also in cunning. She tells Eurükleia to move their bed outside their bedroom and to make it up for Odysseus there. Odysseus's rage at hearing that identifies him beyond doubt, for she knew that only Odysseus would know—it is their "pact and pledge" and "secret sign"—that the bed could not be moved without destroying it. He built their bedroom with his own hands, and an old olive tree, as he says,

> grew like a pillar on the building plot,
> and I laid out our bedroom round that tree . . .
> . . . I lopped off the silvery leaves and branches,
> hewed and shaped that stump from the roots up
> into a bedpost . . .

She acknowledges him then, and only then does she give herself to his embrace.

> Now from his heart into his eyes the ache
> of longing mounted, and he wept at last,
> his dear wife, clear and faithful in his arms,
> longed for
> as the sunwarmed earth is longed for by a swimmer
> spent in rough water where his ship went down . . .

And so in the renewal of his marriage, the return of Odysseus and the restoration of order are complete. The order of the kingdom is centered on the marriage bed of the king and queen, and that bed is rooted in the earth. The figure last quoted makes explicit at last the long-hinted analogy between Odysseus's fidelity to his wife and his fidelity to his homeland. In Penélopê's welcoming embrace his two fidelities become one.

For Odysseus, then, marriage was not merely a legal bond, nor even merely a sacred bond, between himself and Penélopê. It was part of a complex practical circumstance involving, in addition to husband and wife, their family of both descendants and forebears, their household, their community, and the sources of all these lives in memory and tradition, in the countryside, and in the earth. These things, wedded together in his marriage, he thought of as his home, and it held his love and faith so strongly that sleeping with a goddess could not divert or console him in his exile.

In Odysseus's return, then, we see a complete marriage and a complete fidelity. To reduce marriage, as we have done, to a mere contract of sexual exclusiveness is at once to degrade it and to make it impossible. That is to take away its dignity and its potency of joy, and to make it only a pitiful little duty —not a union, but a division and a solitude.

The Odyssey's understanding of marriage as the vital link that joins the human community and the earth is obviously full of political implication. In this it will remind us of the Confucian principle that "the government of the state is rooted in family order." But *The Odyssey* goes further than the Confucian texts, it seems to me, in its understanding of agricultural value as the foundation of domestic order and peace.

I have considered the poem so far as describing a journey from the nonhuman order of the sea wilderness to the human order of the cleansed and reunited household. But it is also a journey between two kinds of human value; it moves from the battlefield of Troy to the terraced fields of Ithaka, which, through all the years and great deeds of Odysseus's absence, the peasants have not ceased to farm.

The Odyssey begins in the world of *The Iliad*, a world that, like our own, is war-obsessed, preoccupied with "manly" deeds of exploitation, anger, aggression, pillage, and the disorder, uprootedness, and vagabondage that are their result. At the end of the poem, Odysseus moves away from the values of that world toward the values of domesticity and peace. He restores order to his household by an awesome violence, it is true. But that finished and the house purified, he reenters his marriage, the bedchamber, and the marriage bed rooted in the earth. From there he goes into the fields.

The final recognition scene occurs between Odysseus and his old father, Laërtês:

> Odysseus found his father in solitude
> spading the earth around a young fruit tree.
>
> He wore a tunic, patched and soiled, and leggings—
> oxhide patches, bound below his knees
> against the brambles . . .

The point is not stated—the story is moving so evenly now toward its conclusion that it will not trouble to remind us that the man thus dressed is a *king*—but it is clear that Laërtês has survived his son's absence and the consequent grief and disorder *as a peasant*. Although Odysseus jokes about his father's appearance, the appropriateness of what he is doing is never questioned. In a time of disorder he has returned to the care of the earth, the foundation of life and hope. And Odysseus finds him in an act emblematic of the best and most responsible kind of agriculture: an old man caring for a young tree.

But the homecoming of Odysseus is still not complete. During his wanderings, he was instructed by the ghost of the seer Teirêsias to perform what is apparently to be a ritual of atonement. As the poem ends he still has this before him. Carrying an oar on his shoulder, he must walk inland until he comes to a place where men have no knowledge of the sea or ships, where a passerby will mistake his oar for a winnowing fan. There he must "plant" his oar in the ground and make a sacrifice to the sea god, Poseidon. Home again, he must sacrifice to all the gods. Like those people of the Biblical prophecy who will "beat their swords into plowshares, and their spears into pruning hooks" and not "learn war any more," Odysseus will not know rest until he has carried the instrument of his sea wanderings inland and planted it like a tree, until he has seen the symbol of his warrior life as a farming tool. But after his atonement has been made, a gentle death will come to him when he is weary with age, his countrymen around him "in blessed peace."

The Odyssey, then, is in a sense an anti-*Iliad*, posing against the warrior values of the other epic — the glories of battle and foreign adventuring — an affirmation of the values of domesticity and farming. But at the same time *The Odyssey* is too bountiful and wise to set these two kinds of value against each other in any purity or exclusiveness of opposition. Even less does it set into such opposition the two kinds of experience. The point seems to be that these apparently opposed experiences are linked together. The higher value may be given to domesticity, but this cannot be valued or understood alone. Odysseus's fidelity and his homecoming are as moving and instructive as they are precisely because they are the result of *choice*. We know — as Odysseus undoubtedly does also — the extent of his love for Penélopê because he can return to her only by choosing her, at the price of death, over Kalypso. We feel and understand, with Odysseus, the value of Ithaka as a homeland, because bound inextricably to the experience of his return is the memory of his absence, of his long wandering at sea, and even of the excitement of his adventures. The prophecy of the peaceful death that is to come to him is so deeply touching because the poem has so fully realized the experiences of discord and violent death. The farm life of the island seems so sweet and orderly because we know the dark wilderness of natural force and mystery within which the fields are cleared and lighted.

The Necessity of Wildness

Domestic order is obviously threatened by the margin of wilderness that surrounds it. Marriage may be destroyed by instinctive sexuality; the husband may choose to remain with Kalypso or the wife may run away with godlike Paris. And the forest is always waiting to overrun the fields. These are real possibilities. They must be considered, respected, even feared.

And yet I think that no culture that hopes to endure can afford to destroy them or to set up absolute safeguards against them. Invariably the failure of organized religions, by which they cut themselves off from mystery and therefore from sanctity, lies in the attempt to impose an absolute division between faith and doubt, to make belief perform as knowledge; when they forbid their prophets to go into the wilderness, they lose the possibility of renewal. And the most dangerous tendency in modern society, now rapidly emerging as a scientific-industrial ambition, is the tendency toward encapsulation of human order — the severance, once and for all, of the umbilical cord fastening us to the wilderness or the Creation. The threat is not only in the totalitarian desire for absolute control. It lies in the willingness to ignore an essential paradox:

the natural forces that so threaten us are the same forces that preserve and renew us.

An enduring agriculture must never cease to consider and respect and preserve wildness. The farm can exist only within the wilderness of mystery and natural force. And if the farm is to last and remain in health, the wilderness must survive within the farm. That is what agricultural fertility *is:* the survival of natural process in the human order. To learn to preserve the fertility of the farm, Sir Albert Howard wrote, we must study the forest.

Similarly, the instinctive sexuality within which marriage exists must somehow be made to thrive within marriage. To divide one from the other is to degrade both and ultimately to destroy marriage.

Fidelity to human order, then, if it is fully responsible, implies fidelity also to natural order. Fidelity to human order makes devotion possible. Fidelity to natural order preserves the possibility of choice, the possibility of the renewal of devotion. Where there is no possibility of choice, there is no possibility of faith. One who returns home—to one's marriage and household and place in the world—desiring anew what was previously chosen, is neither the world's stranger nor its prisoner, but is at once in place and free.

The relation between these two fidelities, inasmuch as they sometimes appear to contradict each other, cannot help but be complex and tricky. In our present stage of cultural evolution, it cannot help but be baffling as well. And yet it is only the double faith that is adequate to our need. If we are to have a culture as resilient and competent in the face of necessity as it needs to be, then it must somehow involve within itself a ceremonious generosity toward the wilderness of natural force and instinct. The farm must yield a place to the forest, not as a wood lot, or even as a necessary agricultural principle, but as a sacred grove—a place where the Creation is let alone, to serve as instruction, example, refuge; a place for people to go, free of work and presumption, to let themselves alone. And marriage must recognize that it survives because of, as well as in spite of, Kalypso and Paris and the generosity of instinct that they represent. It must give some ceremonially acknowledged place to the sexual energies that now thrive outside all established forms, in the destructive freedom of moral ignorance or disregard. Without these accommodations we will remain divided: some of us will continue to destroy the world for purely human ends, while others, for the sake of nature, will abandon the task of human order.

What forms or revisions of forms may be adequate to this double faith, I do not know. Cultural solutions are organisms, not machines, and they cannot be invented deliberately or imposed by prescription. Perhaps all that one can do is to clarify as well as possible the needs and pressures that bear upon the process of cultural evolution. I am certain, however, that no satisfactory solution can come from considering marriage alone or agriculture alone. These are our basic connections to each other and to the earth, and they tend to relate analogically and to be reciprocally defining: our demands upon the earth are determined by our ways of living with one another; our regard for one another is brought to light in our ways of using the earth. And I am certain that neither can be changed for the better in the experimental, prescriptive ways we have been using. Ways of life change only in living. To live by expert advice is to abandon one's life.

"Freedom" from Fertility

The household is the bond of marriage that is most native to it, that grows with it and gives it substantial being in the world. It is the practical condition within which husband and wife can enact devotion and loyalty to each other. The motive power of sexual love is thus joined directly to constructive work and is given communal and ecological value. Without the particular demands and satisfactions of the making and keeping of a household, the sanctity and legality of marriage remain abstract, in effect theoretical, and its sexuality becomes a danger. Work is the health of love. To last, love must enflesh itself in the materiality of the world—produce food, shelter, warmth or shade, surround itself with careful acts, well-made things. This, I think, is what Millen Brand means in *Local Lives* when he speaks of the "threat" of love—"so that perhaps acres of earth and its stones are needed and drawn-out work and monotony/to balance that danger . . ."

Marriage and the care of the earth are each other's disciplines. Each makes possible the enactment of fidelity toward the other. As the household has become increasingly generalized as a function of the economy and, as a consequence, has become increasingly "mobile" and temporary, these vital connections have been weakened and finally broken. And whatever has been thus disconnected has become a ground of exploitation for some breed of salesman, specialist, or expert.

A direct result of the disintegration of the household is the division of sexuality from fertility and their virtual takeover by specialists. The specialists of human sexuality are the sexual clinicians and the pornographers, both of whom subsist on the increasing possibility of sex between people who neither know nor care about each other. The specialists of human fertility are the evangelists, technicians, and salesmen of birth control, who subsist upon our failure to see any purpose or virtue in sexual discipline. In this, as in our use of every other kind of energy, our inability to contemplate any measure of restraint or forbearance has been ruinous. Here the impulse is characteristically that of the laboratory scientist: to encapsulate sexuality by separating it absolutely from the problems of fertility.

This division occurs, it seems to me, in a profound cultural failure: the loss of any sense that sexuality and fertility might exist together compatibly in this world. We have lost this possibility because we do not understand, because we cannot bear to consider the meaning of restraint.* The sort of restraint I am talking about is illustrated in a recent *National Geographic* article about the people of Hunza in northern Pakistan. The author is a woman, Sabrina Michaud, and she is talking with a Hunza woman in her kitchen:

"'What have you done to have only one child?' she asks me. Her own children range from 12 to 30 years of age, and seem evenly spaced, four to five years apart. 'We leave our husband's bed until each child is weaned,' she explains simply. But this natural means of birth control has declined, and population has soared."

The woman's remark is thus passed over and not returned to; but if I understand the significance of this paragraph, it is of great importance. The decline of "this natural means of birth control" seems to have been contemporaneous with the coming of roads and "progress" and the opening up of a previously isolated country. What is of interest is that in their isolation in arid, narrow valleys surrounded by the stone and ice of the Karakoram Mountains, these people had practiced sexual restraint as a form of birth control. They had neither our statistical expertise nor our doom-prophets of population growth; it just happened that, placed geographically as they were, they lived always in sight of their agricultural or ecological limits, and they made a competent response.

We have been unable to see the difference between this kind of restraint—a cultural response to an understood practical limit—and the obscure, self-hating, self-congratulating Victorian self-restraint, of which our attitudes and technologies of sexual "freedom" are merely the equally obscure other side.

*At the root of this failure is probably another sexual division: the assignment to women of virtually all responsibility for sexual discipline.

This so-called freedom fragments us and turns us more vehemently and violently than before against our own bodies and against the bodies of other people.

For the care or control of fertility, both that of the earth and that of our bodies, we have allowed a technology of chemicals and devices to replace entirely the cultural means of ceremonial forms, disciplines, and restraints. We have gathered up the immense questions that surround the coming of life into the world and reduced them to simple problems for which we have manufactured and marketed simple solutions. An infertile woman and an infertile field both receive a dose of chemicals, at the calculated risk of undesirable consequences, and are thus equally reduced to the status of productive machines. And for unwanted life—sperm, ova, embryos, weeds, insects, etc. —we have the same sort of ready remedies, for sale, of course, and characteristically popularized by advertisements that speak much of advantages but little of problems.

The result is that we are bringing up a generation of young people who feel that they are "free from worry" about fertility. The pharmacist or the doctor will look after the fertility of the body, and the farming experts and agribusinessmen will look after the fertility of the earth. This is to short-circuit human culture at its source. It is, in effect, to remove from consciousness the two fundamental issues of human life. It permits two great powers to be regarded and used as if they were unimportant.

More serious is the resort to "authorized" modes of direct violence. In land use, this is the permanent diminishment or destruction of fertility as an allowable cost of production, as in strip mining or in the sort of agriculture that good farmers have long referred to also as "mining." This use of technological means cuts across all issues of health and culture for the sake of an annual quota of production.

The human analogue is in the "harmless" and "simple" surgeries of permanent sterilization, which are now being promoted by a propaganda of extreme oversimplification. The publicity on this subject is typically evangelical in tone and simplistically moral; the operations are recommended like commercial products by advertisings complete with exuberant testimonials of satisfied customers and appeals to the prospective customer's maturity, sexual pride, and desire for "freedom"; and the possible physical and psychological complications are played down, misrepresented, not mentioned at all, or simply not known. It is altogether possible that the operations will be performed by doctors as perfunctory, simplistic, presumptuous, and uninforming as the public literature.

I am fully aware of the problem of overpopulation, and I do not mean to say that birth control is unnecessary. What I do mean to say is that any means of birth control is a serious matter, both culturally and biologically, and that sterilization is the most serious of all: to give up fertility is a major change, as important as birth, puberty, marriage, or death.

The great changes having to do with a woman's fertility—puberty, childbirth, and menopause—have, like sexual desire, the unarguable sanction of biological determinism. They belong to a kind of natural tradition. As a result, they are not only endurable, but they belong to a process—the life process or the Wheel of Life—that we have learned to affirm with some understanding. We know, among other things, that this process includes tragedy and survives it, even triumphs over it. The same applies to the occasions of a man's fertility, although not so formidably, a man being less involved, physically, in the *predicament* of fertility and consciously involved in it only if he wants to be. Nevertheless, he comes to fertility and, if he is a moral person, to the same issues of responsibility that it poses for women.

One of the fundamental interests of human culture is to impose this responsibility, to subject fertility to moral will. Culture articulates needs and forms for sexual restraint and involves issues of value in the process of mating. It is possible to imagine that the resulting tension creates a distinctly human form of energy, highly productive of works of the hands and the mind. But until recently there was no division between sexuality and fertility, because none was possible.

This division was made possible by modern technology, which subjected human fertility, like the fertility of the earth, to a new kind of will: the technological will, which may not *necessarily* oppose the moral will, but which has not only tended to do so, but has tended to replace it. Simply because it became possible—and simultaneously profitable—we have cut the cultural ties between sexuality and fertility, just as we have cut those between eating and farming. By "freeing" food and sex from worry, we have also set them apart from thought, responsibility, and the issue of quality. The introduction of "chemical additives" has tended to do away with the issue of taste or preference; the specialist of sex, like the specialist of food, is dealing with a commodity, which he can measure but cannot value.

What is horrifying is not only that we are relying so exclusively on a technology of birth control that is still experimental, but that we are using it *casually*, in utter cultural nakedness, unceremoniously, without sufficient understanding, and as a substitute for cultural solutions—exactly as we now employ the technology of land use. And to promote these means without cultural and ecological insight, as merely a way to divorce sexuality from fertility, pleasure from responsibility—or to *sell* them that way for ulterior "moral" motives—is to try to cure a disease by another disease. That is only a new battle in the old war between body and soul—as if we were living in front of a chorus of the most literal fanatics chanting: "If thy right eye offend thee, pluck it out! If thy right hand offend thee, cut it off!"

The technologists of fertility exercise the powers of gods and the social function of priests without community ties or cultural responsibilities. The clinicians of sex change the lives of people—as the clinicians of agriculture change the lives of places and communities—to whom they are strangers and whom they do not know. These specialists thrive in a profound cultural rift, and they are always accompanied by the exploiters who mine that rift for gold. The pornographer exploits sexual division. And working the similar division between us and our land we have the "agribusinessmen," the pornographers of agriculture.

Fertility as Waste

But there is yet another and more direct way in which the isolation of the body has serious agricultural effects. That is in our society's extreme oversimplification of the relation between the body and its food. By regarding it as merely a consumer of food, we reduce the function of the body to that of a conduit which channels the nutrients of the earth from the supermarket to the sewer. Or we make it a little factory that transforms fertility into pollution—to the enormous profit of "agribusiness" and to the impoverishment of the earth. This is another technological and economic interruption of the cycle of fertility.

Much has already been said here about the division between the body and its food in the productive phase of the cycle. It is the alleged wonder of the Modern World that so many people take energy from food in which they have invested no energy, or very little. Ninety-five percent of our people, boasted the former deputy assistant secretary of agriculture, are now free of the "drudgery" of food production. The meanings of that division, as I have been trying to show, are intricate and degenerative. But that is only half of it. Ninety-five percent (at least) of our people are also free of any involvement or interest in the maintenance phase of the cycle. As their bodies take in and use the nutrients of the soil, those nutrients are transformed into what we are pleased to regard as "wastes"—and are duly wasted.

This waste also has its cause in the old "religious" division between body and soul, by which the body and its products are judged offensive. Once, living with this offensiveness was considered a condemnation, and that was bad enough. But modern technology "saved" us with the flush toilet and the waterborne sewage system. These devices deal with the "wastes" of our bodies by simply removing them from consideration. The irony is that this technological purification of the body requires the pollution of the rivers and the starvation of the fields. It makes the alleged offensiveness of the body truly and inescapably offensive and blinds an entire society to the knowledge that these "offensive wastes" are readily purified in the topsoil—that, indeed, from an ecological point of view, these are not wastes and are not offensive, but are valuable agricultural products essential both to the health of the land and to that of the "consumers."

Our system of agriculture, by modeling itself on economics rather than biology, thus removes food from the *cycle* of its production and puts it into a finite, linear process that in effect destroys it by transforming it into waste. That is, it transforms food into fuel, a form of energy that is usable only once, and in doing so it transforms the body into a consumptive machine.

It is strange, but only apparently so, that this system of agriculture is institutionalized, not in any form of rural life or culture, but in what we call our "urban civilization." The cities subsist in competition with the country; they live upon a one-way movement of energies out of the countryside—food and fuel, manufacturing materials, human labor, intelligence, and talent. Very little of this energy is ever returned. Instead of gathering these energies up into coherence, a cultural consummation that would not only return to the countryside what belongs to it, but also give back generosities of learning and art, conviviality and order, the modern city dissipates and wastes them. Along with its glittering "consumer goods," the modern city produces an equally characteristic outpouring of garbage and pollution—just as it produces and/or collects unemployed, unemployable, and otherwise wasted people.

Once again it must be asked, if competition is the appropriate relationship, then why, after generations of this inpouring of rural wealth, materials, and humanity into the cities, are the cities and the countryside in equal states of disintegration and disrepair? Why have the rural and urban communities *both* fallen to pieces?

Health and Work

The modern urban-industrial society is based on a series of radical disconnections between body and soul, husband and wife, marriage and community, community and the earth. At each of these points of disconnection the collaboration of corporation, government, and expert sets up a profit-making enterprise that results in the further dismemberment and impoverishment of the Creation.

Together, these disconnections add up to a condition of critical ill health, which we suffer in common—not just with each other, but with all other creatures. Our economy is based upon this disease. Its aim is to separate us as far as possible from the sources of life (material, social, and spiritual), to put these sources under the control of corporations and specialized professionals, and to sell them to us at the highest profit. It fragments the Creation and sets the fragments into conflict with one another. For the relief of the suffering that comes of this fragmentation and conflict, our economy proposes, not health, but vast "cures" that further centralize power and increase profits: wars, wars on crime, wars on poverty, national schemes of medical aid, insurance, immunization, further industrial and economic "growth," etc.; and these, of course, are followed by more regulatory laws and agencies to see that our health is protected, our freedom preserved, and our money well spent. Although there may be some "good intention" in this, there is little honesty and no hope.

Only by restoring the broken connections can we be healed. Connection *is* health. And what our society does its best to disguise from us is how ordinary, how commonly attainable, health is. We lose our health—and create profitable diseases and dependences—by failing to see the direct connections between living and eating, eating and working, working and loving. In gardening, for instance, one works with the body to feed the body. The work, if it is knowledgeable, makes for excellent food. And it makes one hungry. The work thus makes eating both nourishing and joyful, not consumptive, and keeps the eater from getting fat and weak. This is health, wholeness, a source of delight. And such a solution, unlike the typical industrial solution, does not cause new problems.

The "drudgery" of growing one's own food, then, is not drudgery at all. (If we make the growing of food a drudgery, which is what "agribusiness" does make of it, then we also make a drudgery of eating and of living.) It is—in addition to being the appropriate fulfillment of a practical need—a sacrament, as eating is also, by which we enact and understand our oneness with the Creation, the conviviality of one body with all bodies. This is what we learn from the hunting and farming rituals of tribal cultures.

As the connections have been broken by the fragmentation and isolation of work, they can be restored by restoring the wholeness of work. There is work that is isolating, harsh, destructive, specialized or trivialized into meaninglessness. And there is work that is restorative, convivial, dignified and dignifying, and pleasing. Good work is not just the maintenance of connections —as one is now said to work "for a living" or "to support a family"—but the *enactment* of connections. It *is* living, and a way of living; it is not support for a family in the sense of an exterior brace or prop, but is one of the forms and acts of love.

To boast that now "95 percent of the people can be freed from the drudgery of preparing their own food" is possible only to one who cannot distinguish between these kinds of work. The former deputy assistant secretary cannot see work as a vital connection; he can see it only as a trade of time for money, and so of course he believes in doing as little of it as possible, especially if it involves the use of the body. His ideal is apparently the same as that of a real-estate agency which promotes a rural subdivision by advertising "A homelife of endless vacation." But the society that is so glad to be free of the drudgery of growing and preparing food also boasts a thriving medical industry to which it is paying $500 per person per year. And that is only the down payment.

We embrace this curious freedom and pay its exorbitant cost because of our hatred of bodily labor. We do not want to work "like a dog" or "like an ox" or "like a horse"—that is, we do not want to use ourselves as beasts. This as much as anything is the cause of our disrespect for farming and our abandonment of it to businessmen and experts. We remember, as we should, that there have been agricultural economies that used people as beasts. But that cannot be remedied, as we have attempted to do, by using people as machines, or by not using them at all.

Perhaps the trouble began when we started using animals disrespectfully: as "beasts"—that is, as if they had no more feeling than a machine. Perhaps the destructiveness of our use of machines was prepared in our willingness to abuse animals. That it was never necessary to abuse animals in order to use them is suggested by a passage in *The Horse in the Furrow*, by George Ewart Evans. He is speaking of how the medieval ox teams were worked at the plow: ". . . the ploughman at the handles, the team of oxen—yoked in pairs or four abreast—and the driver who walked alongside with his goad." And then he says: "It is also worth noting that in the Welsh organization . . . the counterpart of the driver was termed *y geilwad* or the *caller*. He walked *backwards* in front of the oxen singing to them as they worked. Songs were specially composed to suit the rhythm of the oxen's work . . ."

That seems to me to differ radically from our present customary use of any living thing. The oxen were not used as beasts or machines, but as fellow creatures. It may be presumed that this work used people the same way. It is possible, then, to believe that there is a kind of work that does not require abuse or misuse, that does not use anything as a substitute for anything else. We are working well when we use ourselves as the fellow creatures of the plants, animals, materials, and other people we are working with. Such work is unifying, healing. It brings us home from pride and from despair, and places us responsibly within the human estate. It defines us as we are: not too good to work with our bodies, but too good to work poorly or joylessly or selfishly or alone.

The domestic joys, the daily housework or business,
the building of houses — they are not phantasms . . .
they have weight and form and location . . .

Walt Whitman, *To Think of Time*

Men and Women in Search of Common Ground

I am not an authority on men or women or any of the possible connections between them. In sexual matters I am an amateur, in both the ordinary and the literal senses of that word. I speak about them only because I am concerned about them; I am concerned about them only because I am involved in them; I am involved in them, apparently, only because I am a human, a qualification for which I deserve no credit.

I do not believe, moreover, that any individual *can* be an authority on the present subject. The common ground between men and women can only be defined by community authority. Individually, we may desire it and think about it, but we are not going to occupy it if we do not arrive there together.

That we have not arrived there, that we apparently are not very near to doing so, is acknowledged by the title of this symposium ["Men and Women in

Search of Common Ground," a symposium at the Jung Institute of San Francisco]. And that a symposium so entitled should be held acknowledges implicitly that we are not happy in our exile. The specific cause of our unhappiness, I assume, is that relationships between men and women are now too often extremely tentative and temporary, whereas we would like them to be sound and permanent.

Apparently, it is in the nature of all human relationships to aspire to be permanent. To propose temporariness as a goal in such relationships is to bring them under the rule of aims and standards that prevent them from beginning. Neither marriage, nor kinship, nor friendship, nor neighborhood can exist with a life expectancy that is merely convenient.

To see that such connections aspire to permanence, we do not have to look farther than popular songs, in which people still speak of loving each other "forever." We now understand, of course, that in this circumstance the word "forever" is not to be trusted. It may mean only "for a few years" or "for a while" or even "until tomorrow morning." And we should not be surprised to realize that if the word "forever" cannot be trusted in this circumstance, then the word "love" cannot be trusted either.

This, as we know, was often true before our own time, though in our time it seems easier than before to say "I will love you forever" and to mean nothing by it. It is possible for such words to be used cynically—that is, they may be *intended* to mean nothing—but I doubt that they are often used with such simple hypocrisy. People continue to use them, I think, because they continue to try to mean them. They continue to express their sexual feelings with words such as "love" and "forever" because they want those feelings to have a transferable value, like good words or good money. They cannot bear for sex to be "just sex," any more than they can bear for family life to be just reproduction or for friendship to be just a mutually convenient exchange of goods and services.

The questions that I want to address here, then, are: Why are sexual and other human relationships now so impermanent? And under what conditions might they become permanent?

* * *

It cannot be without significance that this division is occurring at a time when division has become our characteristic mode of thinking and acting. Everywhere we look now, the axework of division is going on. We see ourselves more and more as divided from each other, from nature, and from what our traditions define as human nature. The world is now full of nations, races, interests, groups, and movements of all sorts, most of them unable to define their relations to each other except in terms of division and opposition. The poor human body itself has been conceptually hacked to pieces and parceled out like a bureaucracy. Brain and brawn, left brain and right brain, stomach, hands, heart, and genitals have all been set up in competition against each other, each supported by its standing army of advocates, press agents, and merchants. In such a time, it is not surprising that the stresses that naturally, and perhaps desirably, occur between the sexes should result in the same sort of division with the same sort of doctrinal justification.

This condition of division is one that we suffer from and complain about, yet it is a condition that we promote by our ambitions and desires and justify by our jargon of "self-fulfillment." Each of us, we say, is supposed to "realize his or her full potential as an individual." It is as if the whole two hundred million of us were saying with Coriolanus:

> I'll never
> Be such a gosling to obey instinct, but stand
> As if a man were author of himself
> And knew no other kin.
>
> (V, iii, 34–37)

By "instinct" he means the love of family, community, and country. In Shakespeare's time, this "instinct" was understood to be the human norm — the definition of humanity, or a large part of that definition. When Coriolanus speaks these lines, he identifies himself, not as "odd," but as monstrous, a *danger* to family, community, and country. He identifies himself, that is, as an individual prepared to act alone and without the restraint of reverence, fidelity, or love. Shakespeare is at one with his tradition in understanding that such a person acted inevitably, not as the "author of himself," but as the author of tragic consequences both for himself and for other people.

The problem, of course, is that we are *not* the authors of ourselves. That we are not is a religious perception, but it is also a biological and a social one. Each of us has had many authors, and each of us is engaged, for better or worse, in that same authorship. We could say that the human race is a great coauthorship in which we are collaborating with God and nature in the making of ourselves and one another. From this there is no escape. We may collaborate either well or poorly, or we may refuse to collaborate, but even to refuse to collaborate is to exert an influence and to affect the quality of the product. This is only a way of saying that by ourselves we have no meaning and no dignity; by ourselves we are outside the human definition, outside our identity. "More and more," Mary Catharine Bateson wrote in *With a Daughter's Eye*, "it has seemed to me that the idea of an individual, the idea that there is someone to be known, separate from the relationships, is simply an error."

* * *

Some time ago I was with Wes Jackson, wandering among the experimental plots at his home and workplace, the Land Institute in Salina, Kansas. We stopped by one plot that had been planted in various densities of population. Wes pointed to a Maximilian sunflower growing alone, apart from the others, and said, "There is a plant that has 'realized its full potential as an individual.'" And clearly it had: It had grown very tall; it had put out many long branches heavily laden with blossoms—and the branches had broken off, for they had grown too long and too heavy. The plant had indeed realized its full potential as an individual, but it had failed as a Maximilian sunflower. We could say that its full potential as an individual *was* this failure. It had failed because it had lived outside an important part of its definition, which consists of *both* its individuality and its community. A part of its properly realizable potential lay in its community, not in itself.

In making a metaphor of this sunflower, I do not mean to deny the value or the virtue of a *proper* degree of independence in the character and economy of an individual, nor do I mean to deny the conflicts that occur between individuals and communities. Those conflicts belong to our definition, too, and are probably as necessary as they are troublesome. I do mean to say that the conflicts are not everything, and that to make conflict—the so-called "jungle law"—the basis of social or economic doctrine is extremely dangerous. A part of our definition is our common ground, and a part of it is sharing and mutually enjoying our common ground. Undoubtedly, also, since we are humans, a part of our definition is a recurring contest over the common ground: Who shall describe its boundaries, occupy it, use it, or own it? But such contests obviously can be carried too far, so that they become destructive both of the commonality of the common ground and of the ground itself.

* * *

The danger of the phrase "common ground" is that it is likely to be meant as no more than a metaphor. I am *not* using it as a metaphor; I mean by it the actual ground that is shared by whatever group we may be talking about—the human race, a nation, a community, or a household. If we use the term only as a metaphor, then our thinking will not be robustly circumstantial and historical, as it needs to be, but only a weak, clear broth of ideas and feelings.

Marriage, for example, is talked about most of the time as if it were only a "human relationship" between a wife and a husband. A good marriage is likely to be explained as the result of mutually satisfactory adjustments of thoughts and feelings—a "deep" and complicated mental condition. That is surely true for some couples some of the time, but, as a general understanding of marriage, it is inadequate and probably unworkable. It is far too much a thing of the mind and, for that reason, is not to be trusted. "God guard me," Yeats wrote, "from those thoughts men think/In the mind alone . . ."

Yeats, who took seriously the principle of incarnation, elaborated this idea in his essay on the Japanese Noh plays, in which he says that "we only believe in those thoughts which have been conceived not in the brain but in the whole body." But we need a broader concept yet, for a marriage involves more than just the bodies and minds of a man and a woman. It involves locality, human circumstance, and duration. There is a strong possibility that the basic human sexual unit is composed of a man and a woman (bodies and minds), plus their history together, plus their kin and descendants, plus their place in the world with its economy and history, plus their natural neighborhood, plus their human community with its memories, satisfactions, expectations, and hopes.

By describing it in such a way, we begin to understand marriage as the insistently practical union that it is. We begin to understand it, that is, as it is represented in the traditional marriage ceremony, those vows being only a more circumstantial and practical way of saying what the popular songs say dreamily and easily: "I will love you forever"—a statement that, in this world, inescapably leads to practical requirements and consequences because it proposes survival as a goal. Indeed, marriage is a union much more than practical, for it looks both to our survival as a species and to the survival of our definition as human beings—that is, as creatures who make promises and keep them, who care devotedly and faithfully for one another, who care properly for the gifts of life in this world.

The business of humanity is undoubtedly survival in this complex sense—a necessary, difficult, and entirely fascinating job of work. We have in us deeply planted instructions—personal, cultural, and natural—to survive, and we do not need much experience to inform us that we cannot survive alone. The smallest possible "survival unit," indeed, appears to be the universe. At any rate, the ability of an organism to survive outside the universe has yet to be demonstrated. Inside it, everything happens *in concert*; not a breath is drawn but by the grace of an inconceivable series of vital connections joining an inconceivable multiplicity of created things in an inconceivable unity. But of course it is preposterous for a mere individual human to espouse the universe—a possibility that is purely mental, and productive of nothing but talk. On the other hand, it may be that our marriages, kinships, friendships, neighborhoods, and all our forms and acts of homemaking are the rites by which we solemnize and enact our union with the universe. These ways are practical, proper, available to everybody, and they can provide for the safekeeping of the small acreages of the universe that have been entrusted to us. Moreover, they give the word "love" its only chance to mean, for only they can give it a history, a community, and a place. Only in such ways can love become flesh and do its worldly work. For example, a marriage without a place, a household, has nothing to show for itself. Without a history of some length, it does not know what it means. Without a community to exert a shaping pressure around it, it may explode because of the pressure inside it.

* * *

These ways of marriage, kinship, friendship, and neighborhood surround us with forbiddings; they are forms of bondage, and involved in our humanity is always the wish to escape. We may be obliged to look on this wish as necessary, for, as I have just implied, these unions are partly shaped by internal pressure. But involved in our humanity also is the warning that we can escape only into loneliness and meaninglessness. Our choice may be between a small, human-sized meaning and a vast meaninglessness, or between the freedom of our virtues and the freedom of our vices. It is only in these bonds that our individuality has a use and a worth; it is only to the people who know us, love us, and depend on us that we are indispensable as the persons we uniquely are. In our industrial society, in which people insist so fervently on their value and their freedom "as individuals," individuals are seen more and more as "units" by their governments, employers, and suppliers. They live, that is, under the rule of the interchangeability of parts: what one person can do, another person can do just as well or a newer person can do better. Separate from the relationships, there is nobody to be known; people become, as they say and feel, nobodies.

It is plain that, under the rule of the industrial economy, humans, at least as individuals, are well advanced in a kind of obsolescence. Among those who have achieved even a modest success according to the industrial formula, the human body has been almost entirely replaced by machines and by a shrinking population of manual laborers. For enormous numbers of people now, the only physical activity that they cannot delegate to machines or menials, who will presumably do it more to their satisfaction, is sexual activity. For many, the only necessary physical labor is that of childbirth.

According to the industrial formula, the ideal human residence (from the Latin *residere*, "to sit back" or "remain sitting") is one in which the residers do not work. The house is built, equipped, decorated, and provisioned by other people, by strangers. In it, the married couple practice as few as possible of the disciplines of household or homestead. Their domestic labor consists principally of buying things, putting things away, and throwing things away, but it is understood that it is "best" to have even those jobs done by an "inferior" person, and the ultimate industrial ideal is a "home" in which *everything* would be done by pushing buttons. In such a "home," a married couple are mates, sexually, legally, and socially, but they are not helpmates; they do nothing useful either together or for each other. According to the ideal, work should be done *away* from home. When such spouses say to each other, "I will love you forever," the meaning of their words is seriously impaired by their circumstances; they are speaking in the presence of so little that they have done and made. Their history together is essentially placeless; it has no visible or tangible incarnation. They have only themselves in view.

* * *

In such a circumstance, the obsolescence of the body is inevitable, and this is implicitly acknowledged by the existence of the "physical fitness movement." Back in the era of the body, when women and men were physically useful as well as physically attractive to one another, physical fitness was simply a condition. Little conscious attention was given to it; it was a by-product of useful work. Now an obsessive attention has been fixed upon it. Physical fitness has become extremely mental; once free, it has become expensive, an industry—just as sexual attractiveness, once the result of physical vigor and useful work, has now become an industry. The history of "sexual liberation" has been a history of increasing bondage to corporations.

Now the human mind appears to be following the human body into obsolescence. Increasingly, jobs that once were done by the minds of individual humans are done by computers—and by governments and experts. Dr. William C. DeVries, the current superstar of industrial heart replacement, can blithely assure a reporter that "the general society is not very well informed to make those decisions [as to the imposition of restraints on medical experiments on human patients], and that's why the medical society or the government who has a wider range of view comes in to make those decisions" (Louisville *Courier-Journal*, 3 Feb. 1985). Thus we may benefit from the "miracles" of modern medical science on the condition that we delegate all moral and critical authority in such matters to the doctors and the government. We may save our bodies by losing our minds, just as, according to another set of experts, we may save our minds by forsaking our bodies. Computer thought is exactly the sort that Yeats warned us against; it is made possible by the assumption that thought occurs "in the mind alone" and that the mind, therefore, is an excerptable and isolatable human function, which can be set aside from all else that is human, reduced to pure process, and so imitated by a machine. But in fact we know that the *human* mind is not distinguishable from what it knows and that what it knows comes from or is radically conditioned by its embodied life in this world. A machine, therefore, cannot be a mind or be like a mind; it can only *replace* a mind.

We know, too, that these mechanical substitutions are part of a long-established process. The industrial economy has made its way among us by a process of division, degradation, and then replacement. It is only after we have been divided against each other that work and the products of work can be degraded; it is only after work and its products have been degraded that workers can be replaced by machines. Only when thought has been degraded can a mind be replaced by a machine, or a society of experts, or a government.

It is true, furthermore, that, in this process of industrialization, what is free is invariably replaced by a substitute that is costly. Bodily health as the result of useful work, for instance, is or was free, whereas industrial medicine, which has flourished upon the uselessness of the body, is damagingly and heartlessly expensive. In the time of the usefulness of the body, when the body became useless it died, and death was understood as a kind of healing; industrial medicine looks upon death as a disease that calls for increasingly expensive cures.

Similarly, in preindustrial country towns and city neighborhoods, the people who needed each other lived close to each other. This proximity was free, and it provided many benefits that were either free or comparatively cheap. This simple proximity has been destroyed and replaced by communications and transportation industries that are, again, enormously expensive and destructive, as well as extremely vulnerable to disruption.

* * *

Insofar as we reside in the industrial economy, our obsolescence, both as individuals and as humankind, is fast growing upon us. But we cannot regret or, indeed, even know that this is true without knowing and naming those never-to-be-official institutions that alone have the power to reestablish us in our true estate and identity: marriage, family, household, friendship, neighborhood, community. For these to have an effective existence, they must be located in the world and in time. So located, they have the power to establish us in our human identity because they are not merely institutions in a public, abstract sense, like the organized institutions but are also private conditions. They are the conditions in which a human is complete, body and mind, because completely necessary and needed.

When we live within these human enclosures, we escape the tyrannical doctrine of the interchangeability of parts; in these enclosures, we live as members, each in its own identity necessary to the others. When our spouse or child, friend or neighbor is in need or in trouble, we do not deal with them by means of a computer, for we know that, with them, we must not think without feeling. We do not help them by sending a machine, for we know that, with them, a machine cannot represent us. We know that, when they need us, we must go and offer ourselves, body and mind, as we are. As members, moreover, we are useless and worse than useless to each other if we do not care properly for the ground that is common to us.

It is only in these trying circumstances that human love is given its chance to have meaning, for it is only in these circumstances that it can be borne out in deeds through time—"even," to quote Shakespeare again, "to the edge of doom"—and thus prove itself true by fulfilling its true term.

In these circumstances, in place and in time, the sexes will find their common ground and be somewhat harmoniously rejoined, not by some resolution of conflict and power, but by proving indispensable to each other, as in fact they are.

Delivered as a speech at a conference, "Spirituality and Healing," at Louisville, Kentucky, on October 17, 1994.

Health Is Membership

I

From our constant and increasing concerns about health, you can tell how seriously diseased we are. Health, as we may remember from at least some of the days of our youth, is at once wholeness and a kind of unconsciousness. Disease (dis-ease), on the contrary, makes us conscious not only of the state of our health but of the division of our bodies and our world into parts.

The word "health," in fact, comes from the same Indo-European root as "heal," "whole," and "holy." To be healthy is literally to be whole; to heal is to make whole. I don't think mortal healers should be credited with the power to make holy. But I have no doubt that such healers are properly obliged to acknowledge and respect the holiness embodied in all creatures, or that our healing involves the preservation in us of the spirit and the breath of God.

If we were lucky enough as children to be surrounded by grown-ups who loved us, then our sense of wholeness is not just the sense of completeness in ourselves but also is the sense of belonging to others and to our place; it is an unconscious awareness of community, of having in common. It may be that this double sense of singular integrity and of communal belonging is our personal standard of health for as long as we live. Anyhow, we seem to know instinctively that health is not divided.

Of course, growing up and growing older as fallen creatures in a fallen world can only instruct us painfully in division and disintegration. This is the stuff of consciousness and experience. But if our culture works in us as it should, then we do not age merely into disintegration and division, but that very experience begins our education, leading us into knowledge of wholeness and of holiness. I am describing here the story of Job, of Lazarus, of the lame man at the pool of Bethesda, of Milton's Samson, of King Lear. If our culture works in us as it should, our experience is balanced by education; we are led out of our lonely suffering and are made whole.

In the present age of the world, disintegration and division, isolation and suffering seem to have overwhelmed us. The balance between experience and education has been overthrown; we are lost in experience, and so-called education is leading us nowhere. We have diseases aplenty. As if that were not enough, we are suffering an almost universal hypochondria. Half the energy of the medical industry, one suspects, may now be devoted to "examinations" or "tests"—to see if, though apparently well, we may not be latently or insidiously diseased.

* * *

If you are going to deal with the issue of health in the modern world, you are going to have to deal with much absurdity. It is not clear, for example, why death should increasingly be looked upon as a curable disease, an abnormality, by a society that increasingly looks upon life as insupportably painful and/or meaningless. Even more startling is the realization that the modern medical industry faithfully imitates disease in the way that it isolates us and parcels us out. If, for example, intense and persistent pain causes you to pay attention only to your stomach, then you must leave home, community, and family and go to a sometimes distant clinic or hospital, where you will be cared for by a specialist who will pay attention only to your stomach.

Or consider the announcement by the Associated Press on February 9, 1994, that "the incidence of cancer is up among all ages, and researchers speculated that environmental exposure to cancer-causing substances other than cigarettes may be partly to blame." This bit of news is offered as a surprise, never mind that the environment (so called) has been known to be polluted and toxic for many years. The blame obviously falls on that idiotic term "the environment," which refers to a world that surrounds us but is presumably different from us and distant from us. Our laboratories have proved long ago that cigarette smoke gets inside us, but if "the environment" surrounds us, how does *it* wind up inside us? So much for division as a working principle of health.

This, plainly, is a view of health that is severely reductive. It is, to begin with, almost fanatically individualistic. The body is seen as a defective or potentially defective machine, singular, solitary, and displaced, without love, solace, or pleasure. Its health excludes unhealthy cigarettes but does not exclude unhealthy food, water, and air. One may presumably be healthy in a disintegrated family or community or in a destroyed or poisoned ecosystem.

* * *

So far, I have been implying my beliefs at every turn. Now I had better state them openly.

I take literally the statement in the Gospel of John that God loves the world. I believe that the world was created and approved by love, that it subsists, coheres, and endures by love, and that, insofar as it is redeemable, it can be redeemed only by love. I believe that divine love, incarnate and indwelling in the world, summons the world always toward wholeness, which ultimately is reconciliation and atonement with God.

I believe that health is wholeness. For many years I have returned again and again to the work of the English agriculturist Sir Albert Howard, who said, in *The Soil and Health*, that "the whole problem of health in soil, plant, animal, and man [is] one great subject."

I am moreover a Luddite, in what I take to be the true and appropriate sense. I am not "against technology" so much as I am for community. When the choice is between the health of a community and technological innovation, I choose the health of the community. I would unhesitatingly destroy a machine before I would allow the machine to destroy my community.

I believe that the community—in the fullest sense: a place and all its creatures—is the smallest unit of health and that to speak of the health of an isolated individual is a contradiction in terms.

* * *

We speak now of "spirituality and healing" as if the only way to render a proper religious respect to the body is somehow to treat it "spiritually." It could be argued just as appropriately (and perhaps less dangerously) that the way to respect the body fully is to honor fully its materiality. In saying this, I intend no reduction. I do not doubt the reality of the experience and knowledge we call "spiritual" any more than I doubt the reality of so-called physical experience and knowledge; I recognize the rough utility of these terms. But I strongly doubt the advantage, and even the possibility, of separating these two realities.

What I'm arguing against here is not complexity or mystery but dualism. I would like to purge my own mind and language of such terms as "spiritual," "physical," "metaphysical," and "transcendental"—all of which imply that the Creation is divided into "levels" that can readily be peeled apart and judged by human beings. I believe that the Creation is one continuous fabric comprehending simultaneously what we mean by "spirit" and what we mean by "matter."

Our bodies are involved in the world. Their needs and desires and pleasures are physical. Our bodies hunger and thirst, yearn toward other bodies, grow tired and seek rest, rise up rested, eager to exert themselves. All these desires may be satisfied with honor to the body and its maker, but only if much else besides the individual body is brought into consideration. We have long known that individual desires must not be made the standard of their own satisfaction. We must consider the body's manifold connections to other bodies and to the world. The body, "fearfully and wonderfully made," is ultimately mysterious both in itself and in its dependences. Our bodies live, the Bible says, by the spirit and the breath of God, but it does not say how this is so. We are not going to *know* about this.

The distinction between the physical and the spiritual is, I believe, false. A much more valid distinction, and one that we need urgently to learn to make, is that between the organic and the mechanical. To argue this—as I am going to do—puts me in the minority, I know, but it does not make me unique. In *The Idea of a Christian Society*, T. S. Eliot wrote, "We may say that religion, as distinguished from modern paganism, implies a life in conformity with nature. It may be observed that the natural life and the supernatural life have a conformity to each other which neither has with the mechanistic life."

Still, I wonder if our persistent wish to deal spiritually with physical things does not come either from the feeling that physical things are "low" and unworthy or from the fear, especially when speaking of affection, that "physical" will be taken to mean "sexual."

The *New York Review of Books* of February 3, 1994, for example, carried a review of the correspondence of William and Henry James along with a photograph of the two brothers standing together with William's arm around Henry's shoulders. Apropos of this picture, the reviewer, John Bayley, wrote that "their closeness of affection was undoubted and even took on occasion a quasi-physical form." It is Mr. Bayley's qualifier, "quasi-physical," that sticks in one's mind. What can he have meant by it? Is this prurience masquerading as squeamishness, or vice versa? Does Mr. Bayley feel a need to assure his psychologically sophisticated readers that even though these brothers touched one another familiarly, they were not homosexual lovers?

The phrase involves at least some version of the old dualism of spirit and body or mind and body that has caused us so much suffering and trouble and that raises such troubling questions for anybody who is interested in health. If you love your brother and if you and your brother are living creatures, how could your love for him not be physical? Not spiritual or mental only, not "quasi-physical," but physical. How could you not take a simple pleasure in putting your arm around him?

Out of the same dualism comes our confusion about the body's proper involvement in the world. People seriously interested in health will finally have to question our society's long-standing goals of convenience and effortlessness. What is the point of "labor saving" if by making work effortless we make it poor, and if by doing poor work we weaken our bodies and lose conviviality and health?

* * *

We are now pretty clearly involved in a crisis of health, one of the wonders of which is its immense profitability both to those who cause it and to those who propose to cure it. That the illness may prove incurable, except by catastrophe, is suggested by our economic dependence on it. Think, for example, of how readily our solutions become problems and our cures pollutants. To cure one disease, we need another. The causes, of course, are numerous and complicated, but all of them, I think, can be traced back to the old idea that our bodies are not very important except when they give us pleasure (usually, now, to somebody's profit) or when they hurt (now, almost invariably, to somebody's profit).

This dualism inevitably reduces physical reality, and it does so by removing its mystery from it, by dividing it absolutely from what dualistic thinkers have understood as spiritual or mental reality.

A reduction that is merely theoretical might be harmless enough, I suppose, but theories find ways of getting into action. The theory of the relative unimportance of physical reality has put itself into action by means of a metaphor by which the body (along with the world itself) is understood as a machine. According to this metaphor—which is now in constant general use—the human heart, for example, is no longer understood as the center of our emotional life or even as an organ that pumps; it is understood as "a pump," having somewhat the same function as a fuel pump in an automobile.

If the body is a machine for living and working, then it must follow that the mind is a machine for thinking. The "progress" here is the reduction of mind to brain and then of brain to computer. This reduction implies and requires the reduction of knowledge to "information." It requires, in fact, the reduction of everything to numbers and mathematical operations.

This metaphor of the machine bears heavily upon the question of what we mean by health and by healing. The problem is that like any metaphor, it is accurate only in some respects. A girl is only in some respects like a red rose; a heart is only in some respects like a pump. This means that a metaphor must be controlled by a sort of humorous intelligence, always mindful of the exact limits within which the comparison is meaningful. When a metaphor begins to control intelligence, as this one of the machine has done for a long time, then we must look for costly distortions and absurdities.

Of course, the body in most ways is not at all like a machine. Like all living creatures and unlike a machine, the body is not formally self-contained; its boundaries and outlines are not so exactly fixed. The body alone is not, properly speaking, a body. Divided from its sources of air, food, drink, clothing, shelter, and companionship, a body is, properly speaking, a cadaver, whereas a machine by itself, shut down or out of fuel, is still a machine. Merely as an organism (leaving aside issues of mind and spirit) the body lives and moves and has its being, minute by minute, by an interinvolvement with other bodies and other creatures, living and unliving, that is too complex to diagram or describe. It is, moreover, under the influence of thought and feeling. It does not live by "fuel" alone.

A mind, probably, is even less like a computer than a body is like a machine. As far as I am able to understand it, a mind is not even much like a brain. Insofar as it is usable for thought, for the association of thought with feeling, for the association of thoughts and feelings with words, for the connections between words and things, words and acts, thought and memory, a mind seems to be in constant need of reminding. A mind unreminded would be no mind at all. This phenomenon of reminding shows the extensiveness of mind—how intricately it is involved with sensation, emotion, memory, tradition, communal life, known landscapes, and so on. How you could locate a mind within its full extent, among all its subjects and necessities, I don't know, but obviously it cannot be located within a brain or a computer.

To see better what a mind is (or is not), we might consider the difference between what we mean by knowledge and what the computer now requires us to mean by "information." Knowledge refers to the ability to do or say the right thing at the right time; we would not speak of somebody who does the wrong thing at the wrong time as "knowledgeable." People who perform well as musicians, athletes, teachers, or farmers are people of knowledge. And such examples tell us much about the nature of knowledge. Knowledge is formal, and it informs speech and action. It is instantaneous; it is present and available when and where it is needed.

"Information," which once meant that which forms or fashions from within, now means merely "data." However organized this data may be, it is not shapely or formal or in the true sense in-forming. It is not present where it is needed; if you have to "access" it, you don't have it. Whereas knowledge moves and forms acts, information is inert. You cannot imagine a debater or a quarterback or a musician performing by "accessing information." A computer chock full of such information is no more admirable than a head or a book chock full of it.

The difference, then, between information and knowledge is something like the difference between a dictionary and somebody's language.

Where the art and science of healing are concerned, the machine metaphor works to enforce a division that falsifies the process of healing because it falsifies the nature of the creature needing to be healed. If the body is a machine, then its diseases can be healed by a sort of mechanical tinkering, without reference to anything outside the body itself. This applies, with obvious differences, to the mind; people are assumed to be individually sane or insane. And so we return to the utter anomaly of a creature that is healthy within itself.

* * *

The modern hospital, where most of us receive our strictest lessons in the nature of industrial medicine, undoubtedly does well at surgery and other procedures that permit the body and its parts to be treated as separate things. But when you try to think of it as a place of healing—of reconnecting and making whole—then the hospital reveals the disarray of the medical industry's thinking about health.

In healing, the body is restored to itself. It begins to live again by its own powers and instincts, to the extent that it can do so. To the extent that it can do so, it goes free of drugs and mechanical helps. Its appetites return. It relishes food and rest. The patient is restored to family and friends, home and community and work.

This process has a certain naturalness and inevitability, like that by which a child grows up, but industrial medicine seems to grasp it only tentatively and awkwardly. For example, any ordinary person would assume that a place of healing would put a premium upon rest, but hospitals are notoriously difficult to sleep in. They are noisy all night, and the routine interventions go on relentlessly. The body is treated as a machine that does not need to rest.

You would think also that a place dedicated to healing and health would make much of food. But here is where the disconnections of the industrial system and the displacement of industrial humanity are most radical. Sir Albert Howard saw accurately that the issue of human health is inseparable from the health of the soil, and he saw too that we humans must responsibly occupy our place in the cycle of birth, growth, maturity, death, and decay, which is the health of the world. Aside from our own mortal involvement, food is our fundamental connection to that cycle. But probably most of the complaints you hear about hospitals have to do with the food, which, according to the testimony I have heard, tends to range from unappetizing to sickening. Food is treated as another unpleasant substance to inject. And this is a shame. For in addition to the obvious nutritional link between food and health, food can be a pleasure. People who are sick are often troubled or depressed, and mealtimes offer three opportunities a day when patients could easily be offered something to look forward to. Nothing is more pleasing or heartening than a plate of nourishing, tasty, beautiful food artfully and lovingly prepared. Anything less is unhealthy, as well as a desecration.

Why should rest and food and ecological health not be the basic principles of our art and science of healing? Is it because the basic principles already are technology and drugs? Are we confronting some fundamental incompatibility between mechanical efficiency and organic health? I don't know. I only know that sleeping in a hospital is like sleeping in a factory and that the medical industry makes only the most tenuous connection between health and food and no connection between health and the soil. Industrial medicine is as little interested in ecological health as is industrial agriculture.

A further problem, and an equally serious one, is that illness, in addition to being a bodily disaster, is now also an economic disaster. This is so whether or not the patient is insured. It is a disaster for us all, all the time, because we all know that personally or collectively, we cannot continue to pay for cures that continue to get more expensive. The economic disturbance that now inundates the problem of illness may turn out to be the profoundest illness of all. How can we get well if we are worried sick about money?

* * *

I wish it were not the fate of this essay to be filled with questions, but questions now seem the inescapable end of any line of thought about health and healing. Here are several more:

1. Can our present medical industry produce an adequate definition of health? My own guess is that it cannot do so. Like industrial agriculture, industrial medicine has depended increasingly on specialist methodology, mechanical technology, and chemicals; thus, its point of reference has become more and more its own technical prowess and less and less the health of creatures and habitats. I don't expect this problem to be solved in the universities, which have never addressed, much less solved, the problem of health in agriculture. And I don't expect it to be solved by the government.

2. How can cheapness be included in the criteria of medical experimentation and performance? And why has it not been included before now? I believe that the problem here is again that of the medical industry's fixation on specialization, technology, and chemistry. As a result, the modern "health care system" has become a way of marketing industrial products, exactly like modern agriculture, impoverishing those who pay and enriching those who are paid. It is, in other words, an industry such as industries have always been.

3. Why is it that medical strictures and recommendations so often work in favor of food processors and against food producers? Why, for example, do we so strongly favor the pasteurization of milk to health and cleanliness in milk production? (Gene Logsdon correctly says that the motive here "is monopoly, not consumer health.")

4. Why do we so strongly prefer a fat-free or a germ-free diet to a chemical-free diet? Why does the medical industry strenuously oppose the use of tobacco, yet complacently accept the massive use of antibiotics and other drugs in meat animals and of poisons on food crops? How much longer can it cling to the superstition of bodily health in a polluted world?

5. How can adequate medical and health care, including disease prevention, be included in the structure and economy of a community? How, for example, can a community and its doctors be included in the same culture, the same knowledge, and the same fate, so that they will live as fellow citizens, sharers in a common wealth, members of one another?

II

It is clear by now that this essay cannot hope to be complete; the problems are too large and my knowledge too small. What I have to offer is an association of thoughts and questions wandering somewhat at random and somewhat lost within the experience of modern diseases and the often bewildering industry that undertakes to cure them. In my ignorance and bewilderment, I am fairly representative of those who go, or go with loved ones, to doctors' offices and hospitals. What I have written so far comes from my various efforts to make as much sense as I can of that experience. But now I had better turn to the experience itself.

On January 3, 1994, my brother John had a severe heart attack while he was out by himself on his farm, moving a feed trough. He managed to get to the house and telephone a friend, who sent the emergency rescue squad.

The rescue squad and the emergency room staff at a local hospital certainly saved my brother's life. He was later moved to a hospital in Louisville, where a surgeon performed a double-bypass operation on his heart. After three weeks John returned home. He still has a life to live and work to do. He has been restored to himself and to the world.

He and those who love him have a considerable debt to the medical industry, as represented by two hospitals, several doctors and nurses, many drugs and many machines. This is a debt that I cheerfully acknowledge. But I am obliged to say also that my experience of the hospital during John's stay was troubled by much conflict of feeling and a good many unresolved questions, and I know that I am not alone in this.

In the hospital what I will call the world of love meets the world of efficiency—the world, that is, of specialization, machinery, and abstract procedure. Or, rather, I should say that these two worlds come together in the hospital but do not meet. During those weeks when John was in the hospital, it seemed to me that he had come from the world of love and that the family members, neighbors, and friends who at various times were there with him came there to represent that world and to preserve his connection with it. It seemed to me that the hospital was another kind of world altogether.

When I said early in this essay that we live in a world that was created and exists and is redeemable by love, I did not mean to sentimentalize it. For this is also a fallen world. It involves error and disease, ignorance and partiality, sin and death. If this world is a place where we may learn of our involvement in immortal love, as I believe it is, still such learning is only possible here because that love involves us so inescapably in the limits, sufferings, and sorrows of mortality.

* * *

Like divine love, earthly love seeks plenitude; it longs for the full membership to be present and to be joined. Unlike divine love, earthly love does not have the power, the knowledge, or the will to achieve what it longs for. The story of human love on this earth is a story by which this love reveals and even validates itself by its failures to be complete and comprehensive and effective enough. When this love enters a hospital, it brings with it a terrifying history of defeat, but it comes nevertheless confident of itself, for its existence and the power of its longing have been proved over and over again even by its defeat. In the face of illness, the threat of death, and death itself, it insists unabashedly on its own presence, understanding by its persistence through defeat that it is superior to whatever happens.

The world of efficiency ignores both loves, earthly and divine, because by definition it must reduce experience to computation, particularity to abstraction, and mystery to a small comprehensibility. Efficiency, in our present sense of the word, allies itself inevitably with machinery, as Neil Postman demonstrates in his useful book *Technopoly*. "Machines," he says, "eliminate complexity, doubt, and ambiguity. They work swiftly, they are standardized, and they provide us with numbers that you can see and calculate with." To reason, the advantages are obvious, and probably no reasonable person would wish to reject them out of hand.

And yet love obstinately answers that no loved one is standardized. A body, love insists, is neither a spirit nor a machine; it is not a picture, a diagram, a chart, a graph, an anatomy; it is not an explanation; it is not a law. It is precisely and uniquely what it is. It belongs to the world of love, which is a world of living creatures, natural orders and cycles, many small, fragile lights in the dark.

In dealing with problems of agriculture, I had thought much about the difference between creatures and machines. But I had never so clearly understood and felt that difference as when John was in recovery after his heart surgery, when he was attached to many machines and was dependent for breath on a respirator. It was impossible then not to see that the breathing of a machine, like all machine work, is unvarying, an oblivious regularity, whereas the breathing of a creature is ever changing, exquisitely responsive to events both inside and outside the body, to thoughts and emotions. A machine makes breaths as a machine makes buttons, all the same, but every breath of a creature is itself a creature, like no other, inestimably precious.

* * *

Logically, in plenitude some things ought to be expendable. Industrial economics has always believed this: abundance justifies waste. This is one of the dominant superstitions of American history—and of the history of colonialism everywhere. Expendability is also an assumption of the world of efficiency, which is why that world deals so compulsively in percentages of efficacy and safety.

But this sort of logic is absolutely alien to the world of love. To the claim that a certain drug or procedure would save 99 percent of all cancer patients or that a certain pollutant would be safe for 99 percent of a population, love, unembarrassed, would respond, "What about the one percent?"

There is nothing rational or perhaps even defensible about this, but it is nonetheless one of the strongest strands of our religious tradition—it is probably the most essential strand—according to which a shepherd, owning a hundred sheep and having lost one, does not say, "I have saved 99 percent of my sheep," but rather, "I have lost one," and he goes and searches for the one. And if the sheep in that parable may seem to be only a metaphor, then go on to the Gospel of Luke, where the principle is flatly set forth again and where the sparrows stand not for human beings but for all creatures: "Are not five sparrows sold for two farthings, and not one of them is forgotten before God?" And John Donne had in mind a sort of equation and not a mere metaphor when he wrote, "If a clod be washed away by the sea, Europe is the less, as well as if a promontory were, as well as if a manor of thy friend's or of thine own were. Any man's death diminishes me."

It is reassuring to see ecology moving toward a similar idea of the order of things. If an ecosystem loses one of its native species, we now know that we cannot speak of it as itself minus one species. An ecosystem minus one species is a different ecosystem. Just so, each of us is made by—or, one might better say, made as—a set of unique associations with unique persons, places, and things. The world of love does not admit the principle of the interchangeability of parts.

When John was in intensive care after his surgery, his wife, Carol, was standing by his bed, grieving and afraid. Wanting to reassure her, the nurse said, "Nothing is happening to him that doesn't happen to everybody."

And Carol replied, "I'm not everybody's wife."

* * *

In the world of love, things separated by efficiency and specialization strive to come back together. And yet love must confront death, and accept it, and learn from it. Only in confronting death can earthly love learn its true extent, its immortality. Any definition of health that is not silly must include death. The world of love includes death, suffers it, and triumphs over it. The world of efficiency is defeated by death; at death, all its instruments and procedures stop. The world of love continues, and of this grief is the proof.

In the hospital, love cannot forget death. But like love, death is in the hospital but not of it. Like love, fear and grief feel out of place in the hospital. How could they be included in its efficient procedures and mechanisms? Where a clear, small order is fervently maintained, fear and grief bring the threat of large disorder.

And so these two incompatible worlds might also be designated by the terms "amateur" and "professional"— amateur, in the literal sense of lover, one who participates for love; and professional in the modern sense of one who performs highly specialized or technical procedures for pay. The amateur is excluded from the professional "field."

For the amateur, in the hospital or in almost any other encounter with the medical industry, the overriding experience is that of being excluded from knowledge — of being unable, in other words, to make or participate in anything resembling an "informed decision." Of course, whether doctors make informed decisions in the hospital is a matter of debate. For in the hospital even the professionals are involved in experience; experimentation has been left far behind. Experience, as all amateurs know, is not predictable, and in experience there are no replications or "controls"; there is nothing with which to compare the result. Once one decision has been made, we have destroyed the opportunity to know what would have happened if another decision had been made. That is to say that medicine is an exact science until applied; application involves intuition, a sense of probability, "gut feeling," guesswork, and error.

In medicine, as in many modern disciplines, the amateur is divided from the professional by perhaps unbridgeable differences of knowledge and of language. An "informed decision" is really not even imaginable for most medical patients and their families, who have no competent understanding of either the patient's illness or the recommended medical or surgical procedure. Moreover, patients and their families are not likely to know the doctor, the surgeon, or any of the other people on whom the patient's life will depend. In the hospital, amateurs are more than likely to be proceeding entirely upon faith — and this is a peculiar and scary faith, for it must be placed not in a god but in mere people, mere procedures, mere chemicals, and mere machines.

It was only after my brother had been taken into surgery, I think, that the family understood the extremity of this deed of faith. We had decided — or John had decided and we had concurred — on the basis of the best advice available. But once he was separated from us, we felt the burden of our ignorance. We had not known what we were doing, and one of our difficulties now was the feeling that we had utterly given him up to what we did not know. John himself spoke out of this sense of abandonment and helplessness in the intensive care unit, when he said, "I don't know what they're going to do to me or for me or with me."

As we waited and reports came at long intervals from the operating room, other realizations followed. We realized that under the circumstances, we could not be told the truth. We would not know, ever, the worries and surprises that came to the surgeon during his work. We would not know the critical moments or the fears. If the surgeon did any part of his work ineptly or made a mistake, we would not know it. We realized, moreover, that if we were told the truth, we would have no way of knowing that the truth was what it was.

We realized that when the emissaries from the operating room assured us that everything was "normal" or "routine," they were referring to the procedure and not the patient. Even as amateurs—perhaps *because* we were amateurs—we knew that what was happening was not normal or routine for John or for us.

* * *

That these two worlds are so radically divided does not mean that people cannot cross between them. I do not know how an amateur can cross over into the professional world; that does not seem very probable. But that professional people can cross back into the amateur world, I know from much evidence. During John's stay in the hospital there were many moments in which doctors and nurses—especially nurses!—allowed or caused the professional relationship to become a meeting between two human beings, and these moments were invariably moving.

The most moving, to me, happened in the waiting room during John's surgery. From time to time a nurse from the operating room would come in to tell Carol what was happening. Carol, from politeness or bravery or both, always stood to receive the news, which always left us somewhat encouraged and somewhat doubtful. Carol's difficulty was that she had to suffer the ordeal not only as a wife but as one who had been a trained nurse. She knew, from her own education and experience, in how limited a sense open-heart surgery could be said to be normal or routine.

Finally, toward the end of our wait, two nurses came in. The operation, they said, had been a success. They explained again what had been done. And then they said that after the completion of the bypasses, the surgeon had found it necessary to insert a "balloon pump" into the aorta to assist the heart. This possibility had never been mentioned, nobody was prepared for it, and Carol was sorely disappointed and upset. The two young women attempted to reassure her, mainly by repeating things they had already said. And then there was a long moment when they just looked at her. It was such a look as parents sometimes give to a sick or suffering child, when they themselves have begun to need the comfort they are trying to give.

And then one of the nurses said, "Do you need a hug?"

"Yes," Carol said.

And the nurse gave her a hug.

Which brings us to a starting place.

"It all turns on affection now," said Margaret. "Affection. Don't you see?"

E. M. Forster, *Howards End*

From Sex, Economy, Freedom, and Community

III

Much of the modern assault on community life has been conducted within the justification and protection of the idea of freedom. Thus, it is necessary to try to see how the themes of freedom and community have intersected.

The idea of freedom, as Americans understand it, owes its existence to the inevitability that people will disagree. It is a way of guaranteeing to individuals and to political bodies the right to be different from one another. A specifically American freedom began with our wish to assert our differences from England, and its principles were then worked out in the effort to deal with differences among the states. The result is the Bill of Rights, of which the cornerstone is the freedom of speech. This freedom is not only the basic guarantee of political liberty but it also obligates public officials and private citizens alike to acknowledge the inherent dignity and worth of individual people. It exists only as an absolute; if it can be infringed at all, then probably it can be destroyed entirely.

But if it is an absolute, it is a peculiar and troubling one. It is not an absolute in the sense that a law of nature is. It is not absolute even as the moral law is. One person alone can uphold the moral law, but one person alone cannot uphold the freedom of speech. The freedom of speech is a public absolute, and it can remain absolute only so long as a sufficient segment of the public believes that it is and consents to uphold it. It is an absolute that can be destroyed by public opinion. This is where the danger lies. If this freedom is abused and if a sufficient segment of the public becomes sufficiently resentful of the abuses, then the freedom will be revoked. It is a freedom, therefore, that depends directly on responsibility. And so the First Amendment alone is not a sufficient guarantee of the freedom of speech.

As we now speak of it, freedom is almost always understood as a public idea having to do with the liberties of individuals. The public dialogue about freedom almost always has to do with the efforts of one group or another to wrest these individual liberties from the government or to protect them from another group. In this situation, it is inevitable that freedom will be understood as an issue of power. This is perhaps as necessary as it is unavoidable. But power is not the only issue related to freedom.

From another point of view, not necessarily incompatible, freedom has long been understood as the consequence of knowing the truth. When Jesus said to his followers, "Ye shall know the truth, and the truth shall make you free," he was not talking primarily about politics, but the political applicability of the statement has been obvious for a long time, especially to advocates of democracy. According to this line of thought, freedom of speech is necessary to political health and sanity because it permits speech—the public dialogue—to correct itself. Thomas Jefferson had this in mind when he said in his first inaugural address, "If there be any among us who would wish to dissolve this Union or to change its republican form, let them stand undisturbed as monuments of the safety with which error of opinion may be tolerated where reason is left free to combat it." The often-cited "freedom to be wrong" is thus a valid freedom, but it is a poor thing by itself; its validity comes from the recognition that error is real, identifiable as such, dangerous to freedom as to much else, and controvertible. The freedom to be wrong is valid, in other words, because it is the unexcisable other half of the freedom to be right. If freedom is understood as merely the privilege of the unconcerned and uncommitted to muddle about in error, then freedom will certainly destroy itself.

But to define freedom only as a public privilege of private citizens is finally inadequate to the job of protecting freedom. It leaves the issue too public and too private. It fails to provide a circumstance for those private satisfactions and responsibilities without which freedom is both pointless and fragile. Here as elsewhere, we need to interpose between the public and the private interests a third interest: that of the community. When there is no forcible assertion of the interest of community, public freedom becomes a sort of refuge for escapees from the moral law — those who hold that there is, in Mary McGrory's words, "no ethical transgression except an indictable one."

Public laws are meant for a public, and they vary, sometimes radically, according to forms of government. The moral law, which is remarkably consistent from one culture to another, has to do with community life. It tells us how we should treat relatives and neighbors and, by metaphorical extension, strangers. The aim of the moral law is the integrity and longevity of the community, just as the aim of public law is the integrity and longevity of a political body. Sometimes, the identities of community and political body are nearly the same, and in that case public laws are not necessary because there is, strictly speaking, no "public." As I understand the term, *public* means simply all the people, apart from any personal responsibility or belonging. A public building, for example, is a building which everyone may use but to which no one belongs, which belongs to everyone but not to anyone in particular, and for which no one is responsible except "public employees." A community, unlike a public, has to do first of all with belonging; it is a group of people who belong to one another and to their place. We would say, "We belong to our community," but never "We belong to our public."

I don't know when the concept of "the public," in our sense, emerged from the concept of "the people." But I am aware that there have been human situations in which the concept of "the public" was simply unnecessary. It is not quite possible, for example, to think of the Bushmen or the Eskimos as "publics" or of any parts of their homelands as "public places." And in the traditional rural villages of England there was no public place but rather a "common." A public, I suppose, becomes necessary when a political body grows so large as to include several divergent communities.

A public government, with public laws and a public system of justice, founded on democratic suffrage, is in principle a good thing. Ideally, it makes possible a just and peaceable settlement of contentions arising between communities. It also makes it possible for a mistreated member of a community to appeal for justice outside the community. But obviously such a government can fall short of its purpose. When a public government becomes identified with a public economy, a public culture, and public fashions of thought, it can become the tool of a public process of nationalism or "globalization" that is oblivious of local differences and therefore destructive of communities.

"Public" and "community," then, are different—perhaps radically different—concepts that under certain circumstances are compatible but that, in the present economic and technological monoculture, tend to be at odds. A community, when it is alive and well, is centered on the household—the family place and economy—and the household is centered on marriage. A public, when it is working in the best way—that is, as a political body intent on justice—is centered on the individual. Community and public alike, then, are founded on respect—the one on respect for the family, the other on respect for the individual. Both forms of respect are deeply traditional, and they are not fundamentally incompatible. But they are different, and that difference, once it is instituted in general assumptions, can be the source of much damage and much danger.

A household, according to its nature, will seek to protect and prolong its own life, and since it will readily perceive its inability to survive alone, it will seek to join its life to the life of a community. A young person, coming of age in a healthy household and community, will understand her or his life in terms of membership and service. But in a public increasingly disaffected and turned away from community, it is clear that individuals must be increasingly disinclined to identify themselves in such a way.

The individual, unlike the household and the community, always has two ways to turn: she or he may turn either toward the household and the community, to receive membership and to give service, or toward the relatively unconditional life of the public, in which one is free to pursue self-realization, self-aggrandizement, self-interest, self-fulfillment, self-enrichment, self-promotion, and so on. The problem is that—unlike a married couple, a household, or a community—one individual represents no fecundity, no continuity, and no harmony. The individual life implies no standard of behavior or responsibility.

I am indebted to Judith Weissman for the perception that there are two kinds of freedom: the freedom of the community and the freedom of the individual. The freedom of the community is the more fundamental and the more complex. A community confers on its members the freedoms implicit in familiarity, mutual respect, mutual affection, and mutual help; it gives freedom its proper aims; and it prescribes or shows the responsibilities without which no one can be legitimately free, or free for very long. But to confer freedom or any other benefits on its members, a community must also be free from outside pressure or coercion. It must, in other words, be so far as possible the cause of its own changes; it must change in response to its own changing needs and local circumstances, not in response to motives, powers, or fashions coming from elsewhere. The freedom of the individual, by contrast, has been construed customarily as a license to pursue any legal self-interest at large and at will in the domain of public liberties and opportunities.

These two kinds of freedom, so understood, are clearly at odds. In modern times, the dominant freedom has been that of the individual, and Judith Weissman believes—correctly, I think—that this self-centered freedom is still the aim of contemporary liberation movements:

> The liberation of the individual self for fulfillments, discoveries, pleasures, and joys, and the definition of oppression as mental and emotional constraints . . . this combination existing at the heart of Shelley's Romantic radicalism remains basically unchanged in later feminist writers. . . .

Freedom defined strictly as individual freedom tends to see itself as an escape from the constraints of community life—constraints necessarily implied by consideration for the nature of a place; by consideration for the needs and feelings of neighbors; by kindness to strangers; by respect for the privacy, dignity, and propriety of individual lives; by affection for a place, its people, and its nonhuman creatures; and by the duty to teach the young.

But certain liberationist intellectuals are not the only ones who have demanded this sort of freedom. Almost everybody now demands it, as she or he has been taught to do by the schools, by the various forms of public entertainment, and by salespeople, advertisers, and other public representatives of the industrial economy. People are instructed to free themselves of all restrictions, restraints, and scruples in order to fulfill themselves as individuals to the utmost extent that the law allows. Moreover, we treat corporations as "persons"—an abuse of a metaphor if ever there was one!—and allow to them the same liberation from community obligations that we allow to individuals.

But there is a paradox in all this, and it is as cruel as it is obvious: as the emphasis on individual liberty has increased, the liberty and power of most individuals has declined. Most people are now finding that they are free to make very few significant choices. It is becoming steadily harder for ordinary people—the unrich, the unprivileged—to choose a kind of work for which they have a preference, a talent, or a vocation, to choose where they will live, to choose to work (or to live) at home, or even to choose to raise their own children. And most individuals ("liberated" or not) choose to conform not to local ways and conditions but to a rootless and placeless monoculture of commercial expectations and products. We try to be "emotionally self-sufficient" at the same time that we are entirely and helplessly dependent for our "happiness" on an economy that abuses us along with everything else. We want the liberty of divorce from spouses and independence from family and friends, yet we remain indissolubly married to a hundred corporations that regard us at best as captives and at worst as prey. The net result of our much-asserted individualism appears to be that we have become "free" for the sake of not much self-fulfillment at all.

However frustrated, disappointed, and unfulfilled it may be, the pursuit of self-liberation is still the strongest force now operating in our society. It is the dominant purpose not only of those feminists whose individualism troubles Judith Weissman but also of virtually the entire population; it determines the ethics of the professional class; it defines increasingly the ambitions of politicians and other public servants. This purpose is publicly sanctioned and publicly supported, and it operates invariably to the detriment of community life and community values.

All the institutions that "serve the community" are publicly oriented: the schools, governments and government agencies, the professions, the corporations. Even the churches, though they may have community memberships, do not concern themselves with issues of local economy and local ecology on which community health and integrity must depend. Nor do the people in charge of these institutions think of themselves as members of communities. They are itinerant, in fact or in spirit, as their careers require them to be. These various public servants all have tended to impose on the local place and the local people programs, purposes, procedures, technologies, and values that originated elsewhere. Typically, these "services" involve a condescension to and a contempt for local life that are implicit in all the assumptions—woven into the very fabric—of the industrial economy.

A community, especially if it is a rural community, is understood by its public servants as provincial, backward and benighted, unmodern, unprogressive, unlike "us," and therefore in need of whatever changes are proposed for it by outside interests (to the profit of the outside interests). Anyone who thinks of herself or himself as a member of such a community will sooner or later see that the community is under attack morally as well as economically. And this attack masquerades invariably as altruism: the community must be plundered, expropriated, or morally offended for its own good — but its good is invariably defined by the interest of the invader. The community is not asked whether or not it wishes to be changed, or how it wishes to be changed, or what it wishes to be changed into. The community is deemed to be backward and provincial, it is taught to believe and to regret that it is backward and provincial, and it is thereby taught to welcome the purposes of its invaders. . . .

It is certain that communities are destroyed both from within and from without: by internal disaffection and external exploitation. It is certain, too, that there have always been people who have become estranged from their communities for reasons of honest difference or disagreement. But it can be argued that community disintegration typically is begun by an aggression of some sort from the outside and that in modern times the typical aggression has been economic. The destruction of the community begins when its economy is made — not *dependent* (for no community has ever been entirely independent) — but *subject* to a larger external economy. As an example, we could probably do no better than the following account of the destruction of the local wool economy of the parish of Hawkshead in the Lake District of England:

> The . . . reason for the decline of the customary tenant must be sought in the introduction of machinery towards the end of the eighteenth century, which extinguished not only the local spinning and weaving, but was also the death-blow of the local market. Before this time, idleness at a fellside farm was unknown, for clothes and even linen were home-made, and all spare time was occupied by the youths in carding wool, while the girls spun the "garn" with distaff and wheel. . . . The sale of the yarn to the local weavers, and at the local market, brought important profits to the dalesman, so that it not only kept all hands busy, but put money into his pocket. But the introduction of machinery for looms and for spinning, and consequent outside demand for fleeces instead of yarn and woven material, threw idle not only half of the family, but the local hand-weavers, who were no doubt younger sons of the same stock. Thus idleness took the place of thrift and industry among a naturally industrious class, for the sons and daughters of the 'statesmen, often too proud to go out to service, became useless encumbrances on the estates. Then came the improvement in agricultural methods [that is, technological innovations], which the 'statesman could not afford to keep abreast of. . . . What else could take place but that which did? The estates became mortgaged and were sold, and the rich manu-

> facturers, whose villas are on the margin of Windermere, have often enough among their servants the actual descendants of the old 'statesmen, whose manufactures they first usurped and whose estates they afterwards absorbed.

This paragraph sets forth the pattern of industrial exploitation of a locality and a local economy, a pattern that has prevailed for two hundred years. The industrialization of the eastern Kentucky coalfields early in the [twentieth] century, though more violent, followed this pattern exactly. A decentralized, fairly independent local economy was absorbed and destroyed by an aggressive, monetarily powerful outside economy. And like the displaced farmers, spinners, and weavers of Hawkshead, the once-independent mountaineers of eastern Kentucky became the wage-earning servants of those who had dispossessed their parents, sometimes digging the very coal that their families had once owned and had sold for as little per acre as the pittance the companies paid per day. By now, there is hardly a rural neighborhood or town in the United States that has not suffered some version of this process.

The same process is destroying local economies and cultures all over the world. Of Ladakh, for example, Helena Norberg-Hodge writes:

> In the traditional culture, villagers provided for their basic needs without money. They had developed skills that enabled them to grow barley at 12,000 feet and to manage yaks and other animals at even higher elevations. People knew how to build houses with their own hands from the materials of the immediate surroundings. The only thing they actually needed from outside the region was salt, for which they traded. They used money in only a limited way, mainly for luxuries.
>
> Now, suddenly, as part of the international money economy, Ladakhis find themselves ever more dependent—even for vital needs—on a system that is controlled by faraway forces. They are vulnerable to decisions made by people who do not even know that Ladakh exists. . . . For two thousand years in Ladakh, a kilo of barley has been a kilo of barley, but now you cannot be sure of its value.

This, I think, speaks for itself: if you are dependent on people who do not know you, who control the value of your necessities, you are not free, and you are not safe.

The industrial revolution has thus made universal the colonialist principle that has proved to be ruinous beyond measure: the assumption that it is permissible to ruin one place or culture for the sake of another. Thus justified or excused, the industrial economy grows in power and thrives on its damages to local economies, communities, and places. Meanwhile, politicians and bureaucrats measure the economic prosperity of their nations according to the burgeoning wealth of the industrial interests, not according to the success or failure of small local economies or the reduction and often hopeless servitude of local people. The self-congratulation of the industrialists and their political minions has continued unabated to this day. And yet it is a fact that the industrialists of Hawkshead, like all those elsewhere and since, have lived off the public, just as surely as do the despised clients of "welfare." They have lived off a public of industrially destroyed communities, and they have not compensated for this destruction by their ostentatious contributions to the art, culture, and education of the professional class, or by their "charities" to the poor. Nor is this state of things ameliorated by efforts to enable a local population to "participate" in the global economy, by "education" or any other means. It is true that local individuals, depending on their capital, intelligence, cunning, or influence, may be able to "participate" to their apparent advantage, but local communities and places can "participate" only as victims. The global economy does not exist to help the communities and localities of the globe. It exists to siphon the wealth of those communities and places into a few bank accounts. To this economy, democracy and the values of the religious traditions mean absolutely nothing. And those who wish to help communities to survive had better understand that a merely political freedom means little within a totalitarian economy. Ms. Norberg-Hodge says, of the relatively new influence of the global economy on Ladakh:

> Increasingly, people are locked into an economic system that pumps resources out of the periphery into the center—from the nonindustrialized to the industrialized parts of the world, from the countryside to the city, from the poor to the rich. Often, these resources end up back where they came from as commercial products . . . at prices that the poor can no longer afford.

This is an apt description not just of what is happening in Ladakh but of what is happening in my own rural county and in every other rural county in the United States.

The situation in the wool economy of Hawkshead at the end of the eighteenth century was the same as that which, a little later, caused the brief uprising of those workers in England who were called Luddites. These were people who dared to assert that there were needs and values that justly took precedence over industrialization; they were people who rejected the determinism of technological innovation and economic exploitation. In them, the community attempted to speak for itself and defend itself. It happened that Lord Byron's maiden speech in the House of Lords, on February 27, 1812, dealt with the uprising of the Luddites, and this, in part, is what he said:

> By the adoption of one species of [weaving] frame in particular, one man performed the work of many, and the superfluous laborers were thrown out of employment. Yet it is to be observed, that the work thus executed was inferior in quality; not marketable at home, and merely hurried over with a view to exportation. . . . The rejected workmen . . . conceived themselves to be sacrificed to improvements in mechanism. In the foolishness of their hearts they imagined that the maintenance and well-doing of the industrious poor were objects of greater importance than the enrichment of a few individuals by any improvement, in the implements of trade, which threw the workmen out of employment, and rendered the laborer unworthy of his hire.

The Luddites did, in fact, revolt not only against their own economic oppression but also against the poor quality of the machine work that had replaced them. And though they destroyed machinery, they "abstained from bloodshed or violence against living beings, until in 1812 a band of them was shot down by soldiers." Their movement was suppressed by "severe repressive legislation" and by "many hangings and transportations."

The Luddites thus asserted the precedence of community needs over technological innovation and monetary profit, and they were dealt with in a way that seems merely inevitable in the light of subsequent history. In the years since, the only group that I know of that has successfully, so far, made the community the standard of technological innovation has been the Amish. The Amish have differed from the Luddites in that they have not destroyed but merely declined to use the technologies that they perceive as threatening to their community. And this has been possible because the Amish are an agrarian people. The Luddites could not have refused the machinery that they destroyed; the machinery had refused them.

The victory of industrialism over Luddism was thus overwhelming and unconditional; it was undoubtedly the most complete, significant, and lasting victory of modern times. And so one must wonder at the intensity with which any suggestion of Luddism still is feared and hated. To this day, if you say you would be willing to forbid, restrict, or reduce the use of technological devices in order to protect the community—or to protect the good health of nature on which the community depends—you will be called a Luddite, and it will not be a compliment. To say that the community is more important than machines is certainly Christian and certainly democratic, but it is also Luddism and therefore not to be tolerated. . . .

Now, having discussed at some length the honored practice of community destruction by economic invasion, it will be useful to look at an example of the analogous practice of moral invasion. In 1989, Actors Theater of Louisville presented the premiere performance of Arthur Kopit's play *Bone-the-Fish*. The Louisville *Courier-Journal* welcomed Mr. Kopit and his play to town with the headline: "Arthur Kopit plans to offend almost everyone." In the accompanying article, Mr. Kopit is quoted as saying of his play:

> I am immodestly proud that it is written in consistently bad taste. It's about vile people who do vile things. They are totally loathsome, and I love them all. . . . I'm almost positive that it has something to offend everyone.

The writer of the article explains that "Kopit wrote 'Bone-the-Fish' out of a counterculture impulse, as a reaction against a complacency that he finds is corrupting American life."

My interest here is not in the quality or the point of Mr. Kopit's play, which I did not see (because I do not willingly subject myself to offense). I am interested in the article about him and his play merely as an example of the conventionality of the artistic intention to offend—and of the complacency of the public willingness not to be offended but passively to accept offense. Here we see the famous playwright coming from the center of culture to a provincial city, declaring his intention to "offend almost everyone," and here we see the local drama critic deferentially explaining the moral purpose of this intention. But the playwright makes three rather curious assumptions: (1) that the Louisville theater audience may be supposed (without proof) to be complacent and corrupt; (2) that they therefore deserve to be offended; and (3) that being offended will make them less complacent and less corrupt.

That Louisville theatergoers are more complacent and corrupt than theatergoers elsewhere (or than Mr. Kopit) is not an issue, for there is no evidence. That they deserve to be offended is not an issue for the same reason. That anyone's complacency and corruption can be corrected by being offended in a theater is merely a contradiction in terms, for people who are corrupted by complacency are by definition not likely to take offense. People who do take offense will be either fundamentally decent or aggressively corrupt. People who are fundamentally decent do not deserve to be offended and cannot be instructed by offense. People who are aggressively corrupt would perhaps see the offense but would not accept it. Mr. Kopit's preferred audience is therefore one that will applaud his audacity and pay no attention at all to his avowed didactic purpose—and this perhaps explains his love for "vile people."

If one thinks of Louisville merely as a public, there is not much of an issue here. Mr. Kopit's play is free speech, protected by the First Amendment, and that is that. If, however, one thinks of Louisville as a community or even as potentially a community, then the issue is sizable, and it is difficult. A public, as I have already suggested, is a rather odd thing; I can't think of anything else that is like it. A community is another matter, for it exists within a system of analogies or likenesses that clarify and amplify its meaning. A healthy community is like an ecosystem, and it includes—or it makes itself harmoniously a part of—its local ecosystem. It is also like a household; it is the household of its place, and it includes the households of many families, human and nonhuman. And to extend Saint Paul's famous metaphor by only a little, a healthy community is like a body, for its members mutually support and serve one another.

If a community, then, is like a household, what are we to make of the artist whose intention is to offend? Would I welcome into my house any stranger who came, proud of his bad taste, professing his love for vile people and proposing to offend almost everyone? I would not, and I do not know anybody who would. To do so would contradict self-respect and respect for loved ones. By the same token, I cannot see that a community is under any obligation to welcome such a person. The public, so far as I can see, has no right to require a community to submit to or support statements that offend it.

I know that for a century or so many artists and writers have felt it was their duty—a mark of their honesty and courage—to offend their audience. But if the artist has a duty to offend, does not the audience therefore have a duty to be offended? If the public has a duty to protect speech that is offensive to the community, does not the community have the duty to respond, to be offended, and so defend itself against the offense? A community, as a part of a public, has no right to silence publicly protected speech, but it certainly has a right not to listen and to refuse its patronage to speech that it finds offensive. It is remarkable, however, that many writers and artists appear to be unable to accept this obvious and necessary limitation on their public freedom; they seem to think that freedom entitles them not only to be offensive but also to be approved and subsidized by the people whom they have offended.

These people believe, moreover, that any community attempt to remove a book from a reading list in a public school is censorship and a violation of the freedom of speech. The situation here involves what may be a hopeless conflict of freedoms. A teacher in a public school ought to be free to exercise his or her freedom of speech in choosing what books to teach and in deciding what to say about them. (This, to my mind, would certainly include the right to teach that the Bible is the word of God and the right to teach that it is not.) But the families of a community surely must be allowed an equal freedom to determine the education of their children. How free are parents who have no choice but to turn their children over to the influence of whatever the public will prescribe or tolerate? They obviously are not free at all. The only solution is trust between a community and its teachers, who will therefore teach as members of the community—a trust that in a time of community disintegration is perhaps not possible. And so the public presses its invasion deeper and deeper into community life under the justification of a freedom far too simply understood. It is now altogether possible for a teacher who is forbidden to teach the Bible to teach some other book that is not morally acceptable to the community, perhaps in order to improve the community by shocking or offending it. It is therefore possible that the future of community life in this country may depend on private schools and home schooling.

Does my objection to the intention to offend and the idea of improvement by offense mean that I believe it is invariably wrong to offend or that I think community and public life do not need improving? Obviously not. I do not mean at all to slight the issues of honesty and of artistic integrity that are involved. But I would distinguish between the intention to offend and the willingness to risk offending. Honesty and artistic integrity do not require anyone to intend to give offense, though they certainly may cause offense. The intention to offend, it seems to me, identifies the would-be offender as a public person. I cannot imagine anyone who is a member of a community who would purposely or gladly or proudly offend it, though I know very well that honesty might require one to do so.

Here we are verging on a distinction that had better be explicitly made. There is a significant difference between works of art made to be the vital possessions of a community (existing or not) and those made merely as offerings to the public. Some artists, and I am one of them, wish to live and work within a community, or within the hope of community, in a given place. Others wish to live and work outside the claims of any community, and these now appear to be an overwhelming majority. There is a difference between these two kinds of artists but not necessarily a division. The division comes when the public art begins to conventionalize an antipathy to community life and to the moral standards that enable and protect community life, as our public art has now done. Mr. Kopit's expressed eagerness to offend a local audience he does not know is representative of this antipathy. Our public art now communicates a conventional prejudice against old people, history, parental authority, religious faith, sexual discipline, manual work, rural people and rural life, anything local or small or inexpensive. At its worst, it glamorizes or glorifies drugs, promiscuity, pornography, violence, and blasphemy. Any threat to suppress or limit these public expressions will provoke much support for the freedom of speech. I concur in this. But as a community artist, I would like to go beyond my advocacy of the freedom of speech to deplore some of the uses that are made of it, and I wish that more of my fellow artists would do so as well.

I wish that artists and all advocates and beneficiaries of the First Amendment would begin to ask, for instance, how the individual can be liberated by disobeying the moral law, when the community obviously can be liberated only by obeying it. I wish that they would consider the probability that there is a direct relation between the public antipathy to community life and local ("provincial") places and the industrial destruction of communities and places. I wish, furthermore, that they could see that artists who make offensiveness an artistic or didactic procedure are drawing on a moral capital that they may be using up. A public is shockable or offendable only to the extent that it is already uncomplacent and uncorrupt—to the extent, in other words, that it is a community or remembers being one. What happens after the audience becomes used to being shocked and is therefore no longer shockable—as is apparently near to being the case with the television audience? What if offenses become stimulants—either to imitate the offenses or to avenge them? And what is the difference between the artist who wishes to offend the "provincials" and the industrialist or developer who wishes to dispossess them or convert them into a "labor force"?

The idea that people can be improved by being offended will finally have to meet the idea (espoused some of the time by some of the same people) that books, popular songs, movies, television shows, sex videos, and so on are "just fiction" or "just art" and therefore exist "for their own sake" and have no influence. To argue that works of art are "only" fictions or self-expressions and therefore cannot cause bad behavior is to argue also that they cannot cause good behavior. It is, moreover, to make an absolute division between art and life, experience and life, mind and body—a division that is intolerable to anyone who is at all serious about being a human or a member of a community or even a citizen.

Ananda Coomaraswamy, who had exhaustive knowledge of the traditional uses of art, wrote that "the purpose of any art . . . is to teach, to delight, *and above all to move*" (my emphasis). Of course art moves us! To assume otherwise not only contradicts the common assumption of teachers and writers from the earliest times almost until now; it contradicts everybody's experience. A cathedral, to mention only one of the most obvious examples, is a work of art made to cause a movement toward God, and this is in part a physical movement required by the building's structure and symbolism. But all works of any power move us, in both body and mind, from the most exalted music or poetry to the simplest dance tune. In fact, a dance tune is as good an example as a cathedral. An influence is cast over us, and we are moved. If we see that the influence is bad, we may be moved to reject it, but that is a second movement; it occurs only after we have felt the power of the influence. People do not patronize the makers of pornographic films and sex videos because they are dispassionate appreciators of bad art; they do so because they wish to be moved. Perhaps the makers of pornographic films do not care what their products move their patrons to do. But if they do care, they are writing a check on moral capital to which they do not contribute. They trust that people who are moved by their work will not be moved to sexual harassment or child molestation or rape. They are banking heavily on the moral decency of their customers. And so are all of us who defend the freedom of speech. We are trusting—and not comfortably—that people who come under the influence of the sexual pandering, the greed, the commercial seductions, the moral oversimplification, the brutality, and the violence of our modern public arts will yet somehow remain under the influence of Moses and Jesus. I don't see how anyone can extend this trust without opposing in every way short of suppression the abuses and insults that are protected by it. The more a society comes to be divided in its assumptions and values, the more necessary public freedom becomes. But the more necessary public freedom becomes, the more necessary community responsibility becomes. This connection is unrelenting. And we should not forget that the finest works of art make a community of sorts of their audience. They do not divide people or justify or flatter their divisions; they define our commonwealth, and they enlarge it.

The health of a free public—especially that of a large nation under a representative government—depends on distrust. Thomas Jefferson thought so, and I believe he was right. In subscribing, generation after generation, to our Constitution, we extend to one another and to our government a trust that would be foolish if there were any better alternative. It is a breathtaking act of faith. And this trust is always so near to being misapplied that it cannot be maintained without distrust. People would fail it worse than they do if it were not for the constant vigilance and correction of distrust.

But a community makes itself up in more intimate circumstances than a public. And the health of a community depends absolutely on trust. A community knows itself and knows its place in a way that is impossible for a public (a nation, say, or a state). A community does not come together by a covenant, by a conscientious granting of trust. It exists by proximity, by neighborhood; it knows face-to-face, and it trusts as it knows. It learns, in the course of time and experience, what and who can be trusted. It knows that some of its members are untrustworthy, and it can be tolerant, because to know in this matter is to be safe. A community member can be trusted to be untrustworthy and so can be included. (A community can trust its liars to be liars, for example, and so enjoy them.) But if a community withholds trust, it withholds membership. If it cannot trust, it cannot exist.

One of the essential trusts of community life is that which holds marriages and families together. Another trust is that neighbors will help one another. Another is that privacy will be respected, especially the privacy of personal feeling and the privacy of relationships. All these trusts are absolutely essential, and all are somewhat fragile. But the most fragile, the most vulnerable to public invasion, is the trust that protects privacy. And in our time privacy has been the trust that has been most subjected to public invasion.

I am referring not just to the pryings and snoopings of our secret government, which contradict all that our public government claims to stand for, but also to those by now conventional publications of private grief, of violence to strangers, of the sexual coupling of strangers—all of which allow the indulgence of curiosity without sympathy. These all share in the evil of careless or malicious gossip; like careless or malicious gossip of any other kind, they destroy community by destroying respect for personal dignity and by destroying compassion. It is clear that no self-respecting human being or community would tolerate for a moment the representation of brutality or murder, on television or anywhere else, in such a way as to allow no compassion for the victim. But worst of all—and, I believe, involved in all—is the public prostitution of sex in guises of freedom ranging from the clinical to the commercial, from the artistic to the statistical.

One of the boasts of our century is that its artists—not to mention its psychologists, therapists, anthropologists, sociologists, statisticians, and pornographers—have pried open the bedroom door at last and shown us sexual love for what it "really" is. We have, we assume, cracked the shell of sexual privacy.

The resulting implication that the shell is easily cracked disguises the probability that the shell is, in fact, not crackable at all and that what we have seen displayed is not private or intimate sex, not sexual love, but sex reduced, degraded, oversimplified, and misrepresented by the very intention to display it. Sex publicly displayed is public sex. Sex observed is not private or intimate and cannot be.

Could a voyeur conceivably crack the shell? No, for voyeurs are the most handicapped of all the sexual observers; they know only what they see. True intimacy, even assuming that it can be observed, cannot be known by an outsider and cannot be shown. An artist who undertakes to show the most intimate union of lovers—even assuming that the artist is one of the lovers—can only represent what she or he alone thinks it is. The intimacy, the union itself, remains unobserved. One cannot enter into this intimacy and watch it at the same time, any more than the mind can think about itself while it thinks about something else.

Is sexual love, then, not a legitimate subject of the imagination? It is. But the work of the imagination does not require that the shell be cracked. From Homer to Shakespeare, from the Bible to Jane Austen, we have many imaginings of the intimacy and power of sexual love that have respected absolutely its essential privacy and thus have preserved its intimacy and honored its dignity.

The essential and inscrutable privacy of sexual love is the sign both of its mystery and sanctity and of its humorousness. It is mysterious because the couple who are in it are lost in it. It is their profoundest experience of the being of the world and of their being in it and is at the same time an obliviousness to the world. This lostness of people in sexual love tends to be funny to people who are outside it. But having subscribed to the superstition that we have stripped away all privacy—and mystery and sanctity—from sex, we have become oddly humorless about it. Most people, for example, no longer seem to be aware of the absurdity of sexual vanity. Most people apparently see the sexual pretension and posturing of popular singers, athletes, and movie stars as some kind of high achievement, not the laughable inanity that it really is. Sexual arrogance, on the other hand, is not funny. It is dangerous, and there are some signs that our society has begun to recognize the danger. What it has not recognized is that the publication of sexual privacy is not only fraudulent but often also a kind of sexual arrogance, and a dangerous one.

Does this danger mean that any explicit representation of sexual lovemaking is inevitably wrong? It does not. But it means that such representations *can* be wrong and that when they are wrong, they are destructive.

The danger, I would suggest, is not in the representation but in the reductiveness that is the risk of representation and that is involved in most representations. What is so fearfully arrogant and destructive is the implication that what is represented, or representable, is all there is. In the best representations, I think, there would be a stylization or incompleteness that would convey the artist's honest acknowledgment that this is not all.

The best representations are surrounded and imbued with the light of imagination, so that they make one aware, with profound sympathy, of the two lives, not just the two bodies, that are involved; they make one aware also of the difficulty of full and open sexual consent between two people and of the history and the trust that are necessary to make possible that consent. Without such history and trust, sex is brutal, no matter what species is involved.

When sexual lovemaking is shown in art, one can respond intelligently to it by means of a handful of questions: Are the lovers represented as merely "physical" bodies or as two living souls? Does the representation make it possible to see why Eros has been understood not as an instinct or a "drive" but as a god? Are we asked to see this act as existing in and of and for itself or as joined to the great cycle of fertility and mortality? Does it belong to nature and to culture? Can we imagine this sweetness continuing on through the joys and difficulties of homemaking, the births and the upbringing of children, the deaths of parents and friends—through disagreements, hardships, quarrels, aging, and death? Does it encourage us to forget or to remember that "certainly it must some time come to pass that the very gentle Beatrice will die"?

And finally we must ask how the modern representations of lovemaking that we find in movies, books, paintings, sculptures, and on television measure up to the best love scenes that we know. The best love scene that I know is not explicitly sexual. It is the last scene of *The Winter's Tale*, in which Shakespeare brings onto the same stage, into the one light, young love in its astounding beauty, ardor, and hope, and old love with its mortal wrongs astoundingly graced and forgiven.

The relevance of such imagining is urgently practical; it is the propriety or justness that holds art and the world together. To represent sex without this fullness of imagination is to foreshadow the degradation and destruction of all that is not imagined. Just as the ruin of farmers, farming, and farmland may be predicted from a society's failure to imagine food in all its meanings and connections, so the failure to imagine sex in all its power and sanctity is to prepare the ruin of family and community life and of much else. In order to expose the privacy of sex, we have made of it another industrial specialization, leaving it naked not only of clothes and of customary discretions and courtesies but also of all its cultural and natural connections.

There are, we must realize, kinds of nakedness that are significantly and sometimes ominously different from each other. To know this, we have only to study the examples that are before us. There is — and who can ever forget it? — the nakedness of the photographs of prisoners in Hitler's death camps. This is the nakedness of absolute exposure to mechanical politics, politics gravitating toward the unimagining "efficiency" of machinery. I remember also a photograph of a naked small child running terrified down a dirt road in Vietnam, showing the body's absolute exposure to the indifference of air war, the appropriate technology of mechanical politics.

There is also the nakedness in advertising, in the worst kinds of fashionable or commercial art. This is the nakedness of free-market sexuality, the nakedness that is possible only in a society in which price is the only index of worth.

The nakedness of the death camps and of mechanical war denotes an absolute loss of dignity. In advertising, novels, and movies, the nakedness sometimes denotes a very significant and a very dangerous loss of dignity. Where the body has no dignity, where the sanctity of its own mystery and privacy is not recognized by a surrounding and protecting community, there can be no freedom. To destroy the dignity of the body — the dignity of any and every body — is to prepare the way for the enslaver, the rapist, the torturer, the user of cannon fodder. The nakedness or near-nakedness of some tribal peoples (I judge from the photographs that I have seen) is, in contrast, always dignified, and this dignity rests on a trust so complex and comprehensive as to be virtually unimaginable to us. The public nakedness of our own society involves no trust but only an exploitiveness that is inescapably economic and greedy. It is an abandonment of the self to self-exploitation and to exploitation by others.

There is also the nakedness of innocence, as, for example, in Degas's *Seated Bather Drying Herself*, in the Metropolitan Museum of Art, in which the body is shown in the unaware and unregarded coherence and mystery of its own being. This quality D. H. Lawrence saw and celebrated:

> People were bathing and posturing themselves on the beach
> and all was dreary, great robot limbs, robot breasts
> robot voices, robot even the gay umbrellas.
> But a woman, shy and alone, was washing herself under a tap
> and the glimmer of the presence of the gods was like lilies
> and like water-lilies.

Finally, there is the nakedness of sexual candor. However easy or casual nakedness may have been made by public freedom, the nakedness of sexual candor is not possible except within the culturally delineated conditions that establish and maintain trust. And it is utterly private. It can be suggested in art but not represented. Any effort to represent it, I suspect, will inevitably be bogus. What must we do to earn the freedom of being unguardedly and innocently naked to someone? Our own and other cultures suggest that we must do a lot. We must make promises and keep them. We must assume many fearful responsibilities and do much work. We must build the household of trust.

It is the community, not the public, that is the protector of the possibility of this candor, just as it is the protector of other tender, vulnerable, and precious things—the childhood of children, for example, and the fertility of fields. These protections are left to the community, for they can be protected only by affection and by intimate knowledge, which are beyond the capacities of the public and beyond the power of the private citizen.

IV

If the word *community* is to mean or amount to anything, it must refer to a place (in its natural integrity) and its people. It must refer to a placed people. Since there obviously can be no cultural relationship that is uniform between a nation and a continent, "community" must mean a people locally placed and a people, moreover, not too numerous to have a common knowledge of themselves and of their place. Because places differ from one another and because people will differ somewhat according to the characters of their places, if we think of a nation as an assemblage of many communities, we are necessarily thinking of some sort of pluralism.

There is, in fact, a good deal of talk about pluralism these days, but most of it that I have seen is fashionable, superficial, and virtually worthless. It does not foresee or advocate a plurality of settled communities but is only a sort of indifferent charity toward a plurality of aggrieved groups and individuals. It attempts to deal liberally—that is, by the superficial courtesies of tolerance and egalitarianism—with a confusion of claims.

The social and cultural pluralism that some now see as a goal is a public of destroyed communities. Wherever it exists, it is the result of centuries of imperialism. The modern industrial urban centers are "pluralistic" because they are full of refugees from destroyed communities, destroyed community economies, disintegrated local cultures, and ruined local ecosystems. The pluralists who see this state of affairs as some sort of improvement or as the beginning of "global culture" are being historically perverse, as well as politically naive. They wish to regard liberally and tolerantly the diverse, sometimes competing claims and complaints of a rootless society, and yet they continue to tolerate also the ideals and goals of the industrialism that caused the uprooting. They affirm the pluralism of a society formed by the uprooting of cultures at the same time that they regard the fierce self-defense of still-rooted cultures as "fundamentalism," for which they have no tolerance at all. They look with wistful indulgence and envy at the ruined or damaged American Indian cultures so long as those cultures remain passively a part of our plurality, forgetting that these cultures, too, were once "fundamentalist" in their self-defense. And when these cultures again attempt self-defense — when they again assert the inseparability of culture and place — they are opposed by this pluralistic society as self-righteously as ever. The tolerance of this sort of pluralism extends always to the uprooted and passive, never to the rooted and active.

The trouble with the various movements of rights and liberties that have passed among us in the last thirty years is that they have all been too exclusive and so have degenerated too readily into special pleading. They have, separately, asked us to stop exploiting racial minorities or women or nature, and they have been, separately, right to do so. But they have not, separately or together, come to the realization that we live in a society that exploits, first, everything that is not ourselves and then, inevitably, ourselves. To ask, within this general onslaught, that we should honor the dignity of this or that group is to ask that we should swim up a waterfall.

Any group that takes itself, its culture, and its values seriously enough to try to separate, or to remain separate, from the industrial line of march will be, to say the least, unwelcome in the plurality. The tolerance of these doctrinaire pluralists always runs aground on religion. You may be fascinated by religion, you may study it, anthropologize and psychoanalyze about it, collect and catalogue its artifacts, but you had better not believe in it. You may put into "the canon" the holy books of any group, but you had better not think them holy. The shallowness and hypocrisy of this tolerance is exposed by its utter failure to extend itself to the suffering people of Iraq, who are, by the standards of this tolerance, fundamentalist, backward, unprogressive, and in general not like "us."

The problem with this form of pluralism is that it has no authentic standard; its standard simply is what one group or another may want at the moment. Its professed freedom is not that of community life but rather that of a political group acting on the pattern of individualism. To get farther toward a practicable freedom, the group must measure itself and its wants by standards external to itself. I assume that these standards must be both cultural and ecological. If people wish to be free, then they must preserve the culture that makes for political freedom, and they must preserve the health of the world.

There is an insistently practical question that any person and any group seriously interested in freedom must ask: Can land and people be preserved anywhere by means of a culture that is in the usual sense pluralistic? E. M. Forster, writing *Howards End* in the first decade of th[e last] century, doubted that they could. Nothing that has happened in the intervening [ninety]-odd years diminishes that doubt, and much that has happened confirms it.

A culture capable of preserving land and people can be made only within a relatively stable and enduring relationship between a local people and its place. Community cultures made in this way would necessarily differ, and sometimes radically so, from one place to another, because places differ. This is the true and necessary pluralism. There can, I think, be no national policy of pluralism or multiculturalism but only these pluralities of local cultures. And if these cultures are of any value and worthy of any respect, they will not be elective—not determined by mere wishes—but will be formed in response to local nature and local needs.

At present, the rhetoric of racial and cultural pluralism works against the possibility of a pluralism of settled communities, exactly as do the assumptions and the practices of national and global economies. So long as we try to think of ourselves as African Americans or European Americans or Asian Americans, we will never settle anywhere. For an authentic community is made less in reference to who we are than to where we are. I cannot farm my farm as a European American—or as an American, or as a Kentuckian—but only as a person belonging to the place itself. If I am to use it well and live on it authentically, I cannot do so by knowing where my ancestors came from (which, except for one great-grandfather, I do not know and probably can never know); I can do so only by knowing where I am, what the nature of the place permits me to do here, and who and what are here with me. To know these things, I must ask the place. A knowledge of foreign cultures is useful, perhaps indispensable, to me in my effort to settle here, but it cannot tell me where I am.

That there should be peace, commerce, and biological and cultural outcrosses among local cultures is obviously desirable and probably necessary as well. But such a state of things would be radically unlike what is now called pluralism. To start with, a plurality of settled communities could not be preserved by the present-day pluralists' easy assumption that all cultures are equal or of equal value and capable of surviving together by tolerance. The idea of equality is a good one, so long as it means "equality before the law." Beyond that, the idea becomes squishy and sentimental because of manifest inequalities of all kinds. It makes no sense, for example, to equate equality with freedom. The two concepts must be joined precisely and within strict limits if their association is to make any sense at all. Equality, in certain circumstances, is anything but free. If we have equality and nothing else — no compassion, no magnanimity, no courtesy, no sense of mutual obligation and dependence, no imagination — then power and wealth will have their way; brutality will rule. A general and indiscriminate egalitarianism is free-market culture, which, like free-market economics, tends toward a general and destructive uniformity. And tolerance, in association with such egalitarianism, is a way of ignoring the reality of significant differences. If I merely tolerate my neighbors on the assumption that all of us are equal, that means I can take no interest in the question of which ones of us are right and which ones are wrong; it means that I am denying the community the use of my intelligence and judgment; it means that I am not prepared to defer to those whose abilities are superior to mine, or to help those whose condition is worse; it means that I can be as self-centered as I please.

In order to survive, a plurality of true communities would require not egalitarianism and tolerance but knowledge, an understanding of the necessity of local differences, and respect. Respect, I think, always implies imagination — the ability to see one another, across our inevitable differences, as living souls.

People, Land, and Community (1983)

I would like to speak precisely of the connections that join people, land, and community—to describe, for example, the best human use of a problematical hillside farm. In a healthy culture, these connections are complex. The industrial economy breaks them down by oversimplifying them and in the process raises obstacles that make it hard for us to see what the connections are or ought to be. These are mental obstacles, of course, and there appear to be two major ones: the assumption that knowledge (information) can be "sufficient," and the assumption that time and work are short.

These assumptions will be found implicit in a whole set of contemporary beliefs: that the future can be studied and planned for; that limited supplies can be wasted without harm; that good intentions can safeguard the use of nuclear power. A recent newspaper article says, for example, "A congressionally mandated study of the Ogallala Aquifer is finding no great cause for alarm from [sic] its rapidly dropping levels. The director of the . . . study . . . says that even at current rates of pumping, the aquifer can supply the Plains with water for another forty to fifty years. . . . All six states participating in the study . . . are forecasting increased farm yields based on improved technology." Another article speaks of a different technology with the same optimism: "The nation has invested hundreds of billions of dollars in atomic weapons and at the same time has developed the most sophisticated strategies to fine-tune their use to avoid a holocaust. Yet the system that is meant to activate them is the weakest link in the chain. . . . Thus, some have suggested that what may be needed are warning systems for the warning systems."

Always the assumption is that we can first set demons at large, and then, somehow, become smart enough to control them. This is not childishness. It is not even "human weakness." It is a kind of idiocy, but perhaps we will not cope with it and save ourselves until we regain the sense to call it evil.

The trouble, as in our conscious moments we all know, is that we are terrifyingly ignorant. The most learned of us are ignorant. The acquisition of knowledge always involves the revelation of ignorance—almost *is* the revelation of ignorance. Our knowledge of the world instructs us first of all that the world is greater than our knowledge of it. To those who rejoice in the abundance and intricacy of Creation, this is a source of joy, as it is to those who rejoice in freedom. ("The future comes only by surprise," we say, "—thank God!") To those would-be solvers of "the human problem," who hope for knowledge equal to (capable of controlling) the world, it is a source of unremitting defeat and bewilderment. The evidence is overwhelming that knowledge does not solve "the human problem." Indeed, the evidence overwhelmingly suggests—with Genesis—that knowledge *is* the problem. Or perhaps we should say instead that all our problems tend to gather under two questions about knowledge: Having the ability and desire to know, how and what should we learn? And, having learned, how and for what should we use what we know?

One thing we do know, that we dare not forget, is that better solutions than ours have at times been made by people with much less information than we have. We know too, from the study of agriculture, that the same information, tools, and techniques that in one farmer's hands will ruin land, in another's will save and improve it.

This is not a recommendation of ignorance. To know nothing, after all, is no more possible than to know enough. I am only proposing that knowledge, like everything else, has its place, and that we need urgently now to *put* it in its place. If we want to know and cannot help knowing, then let us learn as fully and accurately as we decently can. But let us at the same time abandon our superstitious beliefs about knowledge: that it is ever sufficient; that it can of itself solve problems; that it is intrinsically good; that it can be used objectively or disinterestedly. Let us acknowledge that the objective or disinterested researcher is always on the side that pays best. And let us give up our forlorn pursuit of the "informed decision."

The "informed decision," I suggest, is as fantastical a creature as the "disinterested third party" and the "objective observer." Or it is if by "informed" we mean "supported by sufficient information." A great deal of our public life, and certainly the most expensive part of it, rests on the assumed possibility of decisions so informed. Examination of private life, however, affords no comfort whatsoever to that assumption. It is simply true that we do not and cannot *know* enough to make any important decision.

Of this dilemma we can take marriage as an instance, for as a condition marriage reveals the insufficiency of knowledge, and as an institution it suggests the possibility that decisions can be informed in another way that *is* sufficient, or approximately so. I take it as an axiom that one cannot know enough to get married, any more than one can predict a surprise. The only people who possess information sufficient to their vows are widows and widowers—who do not know enough to *re*marry.

What is not so well understood now as perhaps it used to be is that marriage is made in an inescapable condition of loneliness and ignorance, to which it, or something like it, is the only possible answer. Perhaps this is so hard to understand now because now the most noted solutions are mechanical solutions, which are often exactly suited to mechanical problems. But we are humans—which means that we not only *have* problems but *are* problems. Marriage is not as nicely trimmed to its purpose as a bottle-stopper; it is a not entirely possible solution to a not entirely soluble problem. And this is true of the other human connections. We can commit ourselves fully to anything—a place, a discipline, a life's work, a child, a family, a community, a faith, a friend—only in the same poverty of knowledge, the same ignorance of result, the same self-subordination, the same final forsaking of other possibilities. If we must make these so final commitments without sufficient information, then what *can* inform our decisions?

In spite of the obvious dangers of the word, we must say first that love can inform them. This, of course, though probably necessary, is not safe. What parent, faced with a child who is in love and going to get married, has not been filled with mistrust and fear—and justly so. We who were lovers before we were parents know what a fraudulent justifier love can be. We know that people stay married for different reasons than those for which they get married and that the later reasons will have to be discovered. Which, of course, is not to say that the later reasons may not confirm the earlier ones; it is to say only that the earlier ones must wait for confirmation.

But our decisions can also be informed—our loves both limited and strengthened—by those patterns of value and restraint, principle and expectation, memory, familiarity, and understanding that, inwardly, add up to *character* and, outwardly, to *culture*. Because of these patterns, and only because of them, we are not alone in the bewilderments of the human condition and human love, but have the company and the comfort of the best of our kind, living and dead. These patterns constitute a knowledge far different from the kind I have been talking about. It is a kind of knowledge that includes information, but is never the same as information. Indeed, if we study the paramount documents of our culture, we will see that this second kind of knowledge invariably implies, and often explicitly imposes, limits upon the first kind: some possibilities must not be explored; some things must not be learned. If we want to get safely home, there are certain seductive songs we must not turn aside for, some sacred things we must not meddle with:

> Great captain,
> a fair wind and the honey lights of home
> are all you seek. But anguish lies ahead;
> the god who thunders on the land prepares it . . .
> . . . .
> One narrow strait may take you through his blows:
> denial of yourself, restraint of shipmates.

This theme, of course, is dominant in Biblical tradition, but the theme itself and its modern inversion can be handily understood by a comparison of this speech of Tirêsias to Odysseus in Robert Fitzgerald's Homer with Tennyson's romantic Ulysses who proposes, like a genetic engineer or an atomic scientist,

> To follow knowledge like a sinking star,
> Beyond the utmost bound of human thought.

Obviously unlike Homer's Odysseus, Tennyson's Ulysses is said to come from Dante, and he does resemble Dante's Ulysses pretty exactly—the critical difference being that Dante thought this Ulysses a madman and a fool, and brings down upon his Tennysonian speech to his sailors one of the swiftest anticlimaxes in literature. The real—the human—knowledge is understood as implying and imposing limits, much as marriage does, and these limits are understood to belong necessarily to the definition of a human being.

* * *

In all this talk about marriage I have not forgot that I am supposed to be talking about agriculture. I am going to talk directly about agriculture in a minute, but I want to insist that I have been talking about it indirectly all along, for the analogy between marriage making and farm making, marriage keeping and farm keeping, is nearly exact. I have talked about marriage as a way of talking about farming because marriage, as a human artifact, has been more carefully understood than farming. The analogy between them is so close, for one thing, because they join us to time in nearly the same way. In talking about time, I will begin to talk directly about farming, but as I do so, the reader will be aware, I hope, that I am talking indirectly about marriage.

When people speak with confidence of the longevity of diminishing agricultural sources—as when they speak of their good intentions about nuclear power—they are probably not just being gullible or thoughtless; they are likely to be speaking from belief in several tenets of industrial optimism: that life is long, but time and work are short; that every problem will be solved by a "technological breakthrough" before it enlarges to catastrophe; that *any* problem can be solved in a hurry by large applications of urgent emotion, information, and money. It is regrettable that these assumptions should risk correction by disaster when they could be cheaply and safely overturned by the study of any agriculture that has proved durable.

To the farmer, Emerson said, "The landscape is an armory of powers. . . ." As he meant it, the statement may be true, but the metaphor is ill-chosen, for the powers of a landscape are available to human use in nothing like so simple a way as are the powers of an armory. Or let us say, anyhow, that the preparations needed for the taking up of agricultural powers are more extensive and complex than those usually thought necessary for the taking up of arms. And let us add that the motives are, or ought to be, significantly different.

Arms are taken up in fear and hate, but it has not been uncharacteristic for a farmer's connection to a farm to begin in love. This has not always been so ignorant a love as it sometimes is now; but always, no matter what one's agricultural experience may have been, one's connection to a newly bought farm will begin in love that is more or less ignorant. One loves the place because present appearances recommend it, and because they suggest possibilities irresistibly imaginable. One's head, like a lover's, grows full of visions. One walks over the premises, saying, "If this were mine, I'd make a permanent pasture here; here is where I'd plant an orchard; here is where I'd dig a pond." These visions are the usual stuff of unfulfilled love and induce wakefulness at night.

When one buys the farm and moves there to live, something different begins. Thoughts begin to be translated into acts. Truth begins to intrude with its matter-of-fact. One's work may be defined in part by one's visions, but it is defined in part too by problems, which the work leads to and reveals. And daily life, work, and problems gradually alter the visions. It invariably turns out, I think, that one's first vision of one's place was to some extent an imposition on it. But if one's sight is clear and if one stays on and works well, one's love gradually responds to the place as it really is, and one's visions gradually image possibilities that are really in it. Vision, possibility, work, and life—*all* have changed by mutual correction. Correct discipline, given enough time, gradually removes one's self from one's line of sight. One works to better purpose then and makes fewer mistakes, because at last one sees where one is. Two human possibilities of the highest order thus come within reach: what one wants can become the same as what one has, and one's knowledge can cause respect for what one knows.

"Correct discipline" and "enough time" are inseparable notions. Correct discipline cannot be hurried, for it is both the knowledge of what ought to be done, and the willingness to do it—*all* of it, properly. The good worker will not suppose that good work can be made properly answerable to haste, urgency, or even emergency. But the good worker knows too that after it is done work requires yet more time to prove its worth. One must stay to experience and study and understand the consequences—must understand them by living with them, and then correct them, if necessary, by longer living and more work. It won't do to correct mistakes made in one place by moving to another place, as has been the common fashion in America, or by adding on another place, as is the fashion in any sort of "growth economy." Seen this way, questions about farming become inseparable from questions about propriety of scale. A farm can be too big for a farmer to husband properly or pay proper attention to. Distraction is inimical to correct discipline, and enough time is beyond the reach of anyone who has too much to do. But we must go farther and see that propriety of scale is invariably associated with propriety of another kind: an understanding and acceptance of the human place in the order of Creation—a proper humility. There are some things the arrogant mind does not see; it is blinded by its vision of what it desires. It does not see what is already there; it never sees the forest that precedes the farm or the farm that precedes the shopping center; it will never understand that America was "discovered" by the Indians. It is the properly humbled mind in its proper place that sees truly, because — to give only one reason — it sees details.

And the good farmer understands that further limits are imposed upon haste by nature, which, except for an occasional storm or earthquake, is in no hurry either. In the processes of most concern to agriculture — the building and preserving of fertility — nature is never in a hurry. During the last seventeen years, for example, I have been working at the restoration of a once exhausted hillside. Its scars are now healed over, though still visible, and this year it has provided abundant pasture, more than in any year since we have owned it. But to make it as good as it is now has taken seventeen years. If I had been a millionaire or if my family had been starving, it would still have taken seventeen years. It can be better than it now is, but that will take longer. For it to live fully in its own possibility, as it did before bad use ran it down, may take hundreds of years.

But to think of the human use of a piece of land as continuing through hundreds of years, we must greatly complicate our understanding of agriculture. Let us start a job of farming on a given place — say an initially fertile hillside in the Kentucky River Valley — and construe it through time:

1. To begin using this hillside for agricultural production — pasture or crop — is a matter of a year's work. This is work in the present tense, adequately comprehended by conscious intention and by the first sort of knowledge I talked about — information available to the farmer's memory and built into his methods, tools, and crop and livestock species. Understood in its present tense, the work does not reveal its value except insofar as the superficial marks of craftsmanship may be seen and judged. But excellent workmanship, as with a breaking plow, may prove as damaging as bad workmanship. The work has not revealed its connections to the place or to the worker. These connections are revealed in time.

2. To live on the hillside and use it for a lifetime gives the annual job of work a past and a future. To live on the hillside and use it without diminishing its fertility or wasting it by erosion still requires conscious intention and information, but now we must say *good* intention and *good* (that is, correct) information, resulting in *good* work. And to these we must now add *character:* the sort of knowledge that might properly be called familiarity, and the affections, habits, values, and virtues (conscious and unconscious) that would preserve good care and good work through hard times.

3. For human life to continue on the hillside through successive genera-

tions requires good use, good work, all along. For in any agricultural place that will waste or erode—and all will—bad work does not permit "muddling through"; sooner or later it ends human life. Human continuity is virtually synonymous with good farming, and good farming obviously must outlast the life of any good farmer. For it to do this, in addition to the preceding requirements, we must have *community*. Without community, the good work of a single farmer or a single family will not mean much or last long. For good farming to last, it must occur in a good farming community—that is, a neighborhood of people who know each other, who understand their mutual dependences, and who place a proper value on good farming. In its cultural aspect, the community is an order of memories preserved consciously in instructions, songs, and stories, and both consciously and unconsciously in *ways*. A healthy culture holds preserving knowledge *in place* for a *long* time. That is, the essential wisdom accumulates in the community much as fertility builds in the soil. In both, death becomes potentiality.

* * *

People are joined to the land by work. Land, work, people, and community are all comprehended in the idea of culture. These connections cannot be understood or described by information—so many resources to be transformed by so many workers into so many products for so many consumers—because they are not quantitative. We can understand them only after we acknowledge that they should be harmonious—that a culture must be either shapely and saving or shapeless and destructive. To presume to describe land, work, people, and community by information, by quantities, seems invariably to throw them into competition with one another. Work is then understood to exploit the land, the people to exploit their work, the community to exploit its people. And then instead of land, work, people, and community, we have the industrial categories of resources, labor, management, consumers, and government. We have exchanged harmony for an interminable fuss, and the work of culture for the timed and harried labor of an industrial economy.

But let me bring these notions to the trial of a more particular example.

Wes Jackson and Marty Bender of the Land Institute have recently worked out a comparison between the energy economy of a farm using draft horses for most of its field work and that of an identical farm using tractors. This is a project a generation overdue, of the greatest interest and importance—in short, necessary. And the results will be shocking to those who assume a direct proportion between fossil fuel combustion and human happiness.

These results, however, have not fully explained one fact that Jackson and

Bender had before them at the start of their analysis and that was still running ahead of them at the end: that in the last twenty-five or thirty years, the Old Order Amish, who use horses for farmwork, doubled their population and stayed in farming, whereas in the same period millions of mechanized farmers were driven out. The reason that this is not adequately explained by analysis of the two energy economies, I believe, is that the problem is by its nature beyond the reach of analysis of any kind. The real or whole reason must be impossibly complicated, having to do with nature, culture, religion, family and community life, as well as with agricultural methodology and economics. What I think we are up against is an unresolvable difference between thought and action, thought and life.

What works *poorly* in agriculture—monoculture, for instance, or annual accounting—can be pretty fully explained, because what works poorly is invariably some oversimplifying *thought* that subjugates nature, people, and culture. What works well ultimately defies explanation because it involves an order that in both magnitude and complexity is ultimately incomprehensible.

Here, then, is a prime example of the futility of a dependence on information. We cannot contain what contains us or comprehend what comprehends us. Yeats said that "Man can embody truth but he cannot know it." The part, that is, cannot comprehend the whole, though it can stand for it (and by it). Synecdoche is possible, and its possibility implies the possibility of harmony between part and whole. If we cannot work on the basis of sufficient information, then we have to work on the basis of an understanding of harmony. That, I take it, is what Sir Albert Howard and Wes Jackson mean when they tell us that we must study and emulate on our farms the natural integrities that precede and support agriculture.

The study of Amish agriculture, like the study of *any* durable agriculture, suggests that we live in sequences of patterns that are formally analogous. These sequences are probably hierarchical, at least in the sense that some patterns are more comprehensive than others; they tend to arrange themselves like inter-nesting bowls—though any attempt to represent their order visually will oversimplify it.

And so we must suspect that Amish horse-powered farms work well, not because—or not *just* because—horses are energy-efficient, but because they are living creatures, and therefore fit harmoniously into a pattern of relationships that are necessarily biological, and that rhyme analogically from ecosystem to crop, from field to farmer. In other words, ecosystem, farm, field, crop, horse, farmer, family, and community are in certain critical ways *like* each other. They are, for instance, all related to health and fertility or reproductivity in about the same way. The health and fertility of each involves and is involved in the health and fertility of all.

It goes without saying that tools can be introduced into this agricultural and ecological order without jeopardizing it—but only up to a certain kind, scale, and power. To introduce a tractor into it, as the historical record now seems virtually to prove, is to begin its destruction. The tractor has been so destructive, I think, because it is *unlike* anything else in the agricultural order, and so it breaks the essential harmony. And with the tractor comes dependence on an energy supply that lies not only off the farm but outside agriculture and outside biological cycles and integrities. With the tractor, both farm and farmer become "resources" of the industrial economy, which always exploits its resources.

We would be wrong, of course, to say that anyone who farms with a tractor is a bad farmer. That is not true. What we must say, however, is that once a tractor is introduced into the pattern of a farm, certain necessary restraints and practices, once implicit in technology, must now reside in the character and consciousness of the farmer—at the same time that the economic pressure to cast off restraint and good practice has been greatly increased.

In a society addicted to facts and figures, anyone trying to speak for agricultural *harmony* is inviting trouble. The first trouble is in trying to say what harmony is. It cannot be reduced to facts and figures—though the lack of it can. It is not very visibly a function. Perhaps we can only say what it may be like. It may, for instance, be like sympathetic vibration: "The A string of a violin . . . is designed to vibrate most readily at about 440 vibrations per second: the note A. If that same note is played loudly not on the violin but near it, the violin A string may hum in sympathy." This may have a practical exemplification in the craft of the mud daubers which, as they trowel mud into their nest walls, hum to it, or at it, communicating a vibration that makes it easier to work, thus mastering their material by a kind of song. Perhaps the hum of the mud dauber only activates that anciently perceived likeness between all creatures and the earth of which they are made. For as common wisdom holds, like *speaks to* like. And harmony always involves such specificities of form as in the mud dauber's song and its nest, whereas information accumulates indiscriminately, like noise.

Of course, in the order of creatures, humanity is a special case. Humans, unlike mud daubers, are not naturally involved in harmony. For humans, harmony is always a human product, an artifact, and if they do not know how to make it and choose to make it, then they do not have it. And so I suggest that, for humans, the harmony I am talking about may bear an inescapable likeness to what we know as moral law—or that, for humans, moral law is a significant part of the notation of ecological and agricultural harmony. A great many people seem to have voted for information as a safe substitute for virtue, but this ignores—among much else—the need to prepare humans to live short lives in the face of long work and long time.

Perhaps it is only when we focus our minds on our machines that time seems short. Time is always running out for machines. They shorten our work, in a sense popularly approved, by simplifying it and speeding it up, but our work perishes quickly in them too as they wear out and are discarded. For the living Creation, on the other hand, time is always coming. It is running out for the farm built on the industrial pattern; the industrial farm burns fertility as it burns fuel. For the farm built into the pattern of living things, as an analogue of forest or prairie, time is a bringer of gifts. These gifts may be welcomed and cared for. To some extent they may be expected. Only within strict limits are they the result of human intention and knowledge. They cannot in the usual sense be made. Only in the short term of industrial accounting can they be thought simply earnable. Over the real length of human time, to be earned they must be deserved.

* * *

From this rather wandering excursion I arrive at two conclusions.

The first is that the modern stereotype of an intelligent person is probably wrong. The prototypical modern intelligence seems to be that of the Quiz Kid—a human shape barely discernible in fluff of facts. It is understood that everything must be justified by facts, and facts are offered in justification of *everything*. If it is a fact that soil erosion is now a critical problem in American agriculture, then more facts will indicate that it is not as bad as it *could* be and that Iowa will continue to have topsoil for as long as seventy more years. If facts show that some people are undernourished in America, further facts reveal that we should all be glad we do not live in India. This, of course, is machine thought.

To think better, to think like the best humans, we are probably going to have to learn again to judge a person's intelligence, not by the ability to recite facts, but by the good order or harmoniousness of his or her surroundings. We must suspect that any statistical justification of ugliness and violence is a revelation of stupidity. As an earlier student of agriculture put it: "The intelligent man, however unlearned, may be known by his surroundings, and by the care of his horse, if he is fortunate enough to own one."

My second conclusion is that any public program to preserve land or produce food is hopeless if it does not tend to right the balance between numbers of people and acres of land, and to encourage long-term, stable connections between families and small farms. It could be argued that our nation has never made an effort in this direction that was knowledgeable enough or serious enough. It is certain that no such effort, here, has ever succeeded. The typical American farm is probably sold and remade—often as part of a larger farm—at least every generation. Farms that have been passed to the second generation of the same family are unusual. Farms that have passed to the third generation are rare.

But our crying need is for an agriculture in which the typical farm would be farmed by the third generation of the same family. It would be wrong to try to say exactly what kind of agriculture that would be, but it may be allowable to suggest that certain good possibilities would be enhanced.

The most important of those possibilities would be the lengthening of memory. Previous mistakes, failures, and successes would be remembered. The land would not have to pay the cost of a trial-and-error education for every new owner. A half century or more of the farm's history would be living memory, and its present state of health could be measured against its own past—something exceedingly difficult *outside* of living memory.

A second possibility is that the land would not be overworked to pay for itself at full value with every new owner.

A third possibility would be that, having some confidence in family continuity in place, present owners would have future owners not only in supposition but *in sight* and so would take good care of the land, not for the sake of something so abstract as "the future" or "posterity," but out of particular love for living children and grandchildren.

A fourth possibility is that having the past so immediately in memory, and the future so tangibly in prospect, the human establishment on the land would grow more permanent by the practice of better carpentry and masonry. People who remembered long and well would see the folly of rebuilding their barns every generation or two, and of building new fences every twenty years.

A fifth possibility would be the development of the concept of *enough*.

Only long memory can answer, for a given farm or locality, How much land is enough? How much work is enough? How much livestock and crop production is enough? How much power is enough?

A sixth possibility is that of local culture. Who could say what that would be? As members of a society based on the exploitation of its own temporariness, we probably should not venture a guess. But we can perhaps speak with a little competence of how it would begin. It would not be imported from critically approved cultures elsewhere. It would not come from watching certified classics on television. It would begin in work and love. People at work in communities three generations old would know that their bodies renewed, time and again, the movements of other bodies, living and dead, known and loved, remembered and loved, in the same shops, houses, and fields. That, of course, is a description of a kind of community dance. And such a dance is perhaps the best way we have to describe harmony.

•

Conservation and Local Economy

In our relation to the land, we are ruled by a number of terms and limits set not by anyone's preference but by nature and by human nature:

I. Land that is used will be ruined unless it is properly cared for.

II. Land cannot be properly cared for by people who do not know it intimately, who do not know how to care for it, who are not strongly motivated to care for it, and who cannot afford to care for it.

III. People cannot be adequately motivated to care for land by general principles or by incentives that are merely economic—that is, they won't care for it merely because they think they should or merely because somebody pays them.

IV. People are motivated to care for land to the extent that their interest in it is direct, dependable, and permanent.

V. They will be motivated to care for the land if they can reasonably expect to live on it as long as they live. They will be more strongly motivated if they can reasonably expect that their children and grandchildren will live on it as long as they live. In other words, there must be a mutuality of belonging: they must feel that the land belongs to them, that they belong to it, and that this belonging is a settled and unthreatened fact.

VI. But such belonging must be appropriately limited. This is the indispensable qualification of the idea of land ownership. It is well understood that ownership is an incentive to care. But there is a limit to how much land can be owned before an owner is unable to take proper care of it. The need for attention increases with the intensity of use. But the *quality* of attention decreases as acreage increases.

VII. A nation will destroy its land and therefore itself if it does not foster in every possible way the sort of thrifty, prosperous, permanent rural households and communities that have the desire, the skills, and the means to care properly for the land they are using.

In an age notoriously impatient of restraints, such a list of rules will hardly be welcome, but that these *are* the rules of land use I have no doubt. I am convinced of their authenticity both by common wisdom and by my own experience and observation. The rules exist; the penalties for breaking them are obvious and severe; the failure of land stewardship in this country is the result of a general disregard for all of them.

As proof of this failure, there is no need to recite again the statistics of land ruination. The gullies and other damages are there to be seen. Very little of our land that is being used — for logging, mining, or farming — is being well used. Much of our land has never been well used. Those of us who know what we are looking at know that this is true. And after observing the worsening condition of our land, we have only to raise our eyes a little to see the worsening condition of those who are using the land and who are entrusted with its care. We must accept as a fact that by now, our country (as opposed to our nation) is characteristically in decline. War, depression, inflation, usury, the attitudes of the industrial economy, social and educational fashions — all have taken their toll. For a long time, the news from everywhere in rural America has been almost unrelievedly bad: bankruptcy, foreclosure, depression, suicide, the departure of the young, the loneliness of the old, soil loss, soil degradation, chemical pollution, the loss of genetic and specific diversity, the extinction or threatened extinction of species, the depletion of aquifers, stream degradation, the loss of wilderness, strip mining, clear-cutting, population loss, the loss of supporting economies, the deaths of towns. Rural American communities, economies, and ways of life that in 1945 were thriving and, though imperfect, full of promise for an authentic human settlement of our land are now as effectively destroyed as the Jewish communities of Poland; the means of destruction were not so blatantly evil, but they have proved just as thorough.

The news of rural decline and devastation has been accompanied, to be sure, by a chorus of professional, institutional, and governmental optimists, who continue to insist that all is well, that we are making things worse only as a way of making things better, that farmers who failed are merely "inefficient producers" for whose failure the country is better off, that money and technology will fill the gaps, that government will fill the gaps, that science will soon free us from our regrettable dependence on the soil. We have heard that it is good business and good labor economics to destroy the last remnants of American wilderness. We have heard that the rural population is actually growing because city people are moving to the country and commuters are replacing farmers. We have heard that the rural economy can be repaired by moving the urban economy out into the country and by replacing rural work with work in factories and offices. And all the while the real conditions of the rural land and the rural people have been getting worse.

Of the general condition of the American countryside, my own community will serve well enough as an example. The town of Port Royal, Kentucky, has a population of about one hundred people. The town came into existence as a trading center, serving the farms in a few square miles of hilly country on the west side of the Kentucky River. It has never been much bigger than it is now. But whereas now it is held together by habit or convenience, once it was held together by a complex local economy. In my mother's childhood, in the years before World War I, there were sixteen business and professional enterprises in the town, all serving the town and the surrounding farms. By the time of my own childhood, in the years before World War II, the number had been reduced to twelve, but the town and its tributary landscape were still alive as a community and as an economy. Now, counting the post office, the town has five enterprises, one of which does not serve the local community. There is now no market for farm produce in the town or within forty miles. We no longer have a garage or repair shop of any kind. We have had no doctor for forty years and no school for thirty. Now, as a local economy and therefore as a community, Port Royal is dying.

What does the death of a community, a local economy, cost its members? And what does it cost the country? So far as I know, we have no economists who are interested in such costs. Nevertheless, when you must drive ten or twenty or more miles to reach a doctor or a school or a mechanic or to find parts for farm machinery, the costs exist, and they are increasing. As they increase, they make the economy of every farm and household less tenable.

As people leave the community or, remaining in the place, drop out of the local economy, as the urban-industrial economy more and more usurps the local economy, as the scale and speed of work increase, care declines. As care declines, the natural supports of the human economy and community also decline, for whatever is used is used destructively.

We in Port Royal are part of an agricultural region surrounded by cities that import much of their food from distant places. Though we urgently need crops that can be substituted for tobacco, we produce practically no vegetables or other foods for consumption in our region. Having no local food economy, we produce a less and less diverse food supply for the general market. This condition implies and virtually requires the abuse of our land and our people, and they are abused.

We are also part of a region that is abundantly and diversely forested, and we have no forest economy. We have no local wood products industry. This makes it almost certain that our woodlands and their owners will be abused, and they are abused.

We provide, moreover, a great deal of recreation for our urban neighbors—hunting, fishing, boating, and the like—and we have the capacity to provide more. But for this we receive little or nothing, and sometimes we suffer damage.

In our region, furthermore, there has been no public effort to preserve the least scrap of land in its pristine condition. And the last decade or so of agricultural depression has caused much logging of the few stands of mature forest in private hands. Now, if we want our descendants to know what the original forest was like—that is, to know the original nature of our land—we must start from scratch and grow the necessary examples over the next two or three hundred years.

My part of rural America is, in short, a colony, like every other part of rural America. Almost the whole landscape of this country—from the exhausted cotton fields of the plantation South to the eroding wheatlands of the Palouse, from the strip mines of Appalachia to the clear-cuts of the Pacific slope—is in the power of an absentee economy, once national and now increasingly international, that is without limit in its greed and without mercy in its exploitation of land and people. Between the prosperity of this vast centralizing economy and the prosperity of any local economy or locality, there is now a radical disconnection. The accounting that measures the wealth of corporations, great banks, and national treasuries takes no measure of the civic or economic or natural health of places like Port Royal, Kentucky; Harpster, Ohio; Indianola, Iowa; Matfield Green, Kansas; Wolf Hole, Arizona; or Nevada City, California—and it does not intend to do so.

In 1912, according to William Allen White, "the county in the United States with the largest assessed valuation was Marion County, Kansas. . . . Marion County happened to have a larger per capita of bank deposits than any other American county. . . . Yet no man in Marion County was rated as a millionaire, but the jails and poorhouses were practically empty. The great per capita of wealth was actually distributed among the people who earned it." This, of course, is the realization of that dream that is sometimes called Jeffersonian but is really the dream of the economically oppressed throughout human history. And because this was a rural county, White was not talking just about bank accounts; he was talking about real capital—usable property. That era and that dream are now long past. Now the national economy, which is increasingly a global economy, no longer prospers by the prosperity of the land and people but by their exploitation.

The Civil War made America safe for the moguls of the railroads and of the mineral and timber industries who wanted to be free to exploit the countryside. The work of these industries and their successors is now almost complete. They have dispossessed, disinherited, and moved into the urban economy almost the entire citizenry; they have defaced and plundered the countryside. And now this great corporate enterprise, thoroughly uprooted and internationalized, is moving toward the exploitation of the whole world under the shibboleths of "globalization," "free trade," and "new world order." The proposed revisions in the General Agreement on Tariffs and Trade are intended solely to further this exploitation. The aim is simply and unabashedly to bring every scrap of productive land and every worker on the planet under corporate control.

The voices of the countryside, the voices appealing for respect for the land and for rural community, have simply not been heard in the centers of wealth, power, and knowledge. The centers have decreed that the voice of the countryside shall be that of Snuffy Smith or L'il Abner, and only that voice have they been willing to hear.

"The business of America is business," a prophet of our era too correctly said. Two corollaries are clearly implied: that the business of the American government is to serve, protect, and defend business; and that the business of the American people is to serve the government, which means to serve business. The costs of this state of things are incalculable. To start with, people in great numbers—because of their perception that the government serves not the country or the people but the corporate economy—do not vote. Our leaders, therefore, are now in the curious—and hardly legitimate—position of asking a very substantial number of people to cheer for, pay for, and perhaps die for a government that they have not voted for.

But when the interests of local communities and economies are relentlessly subordinated to the interests of "business," then two further catastrophes inevitably result. First, the people are increasingly estranged from the native wealth, health, knowledge, and pleasure of their country. And, second, the country itself is destroyed.

It is not possible to look at the present condition of our land and people and find support for optimism. We must not fool ourselves. It is altogether conceivable that we may go right along with this business of "business," with our curious religious faith in technological progress, with our glorification of our own greed and violence always rationalized by our indignation at the greed and violence of others, until our land, our world, and ourselves are utterly destroyed. We know from history that massive human failure is possible. It is foolish to assume that we will save ourselves from any fate that we have made possible simply because we have the conceit to call ourselves *Homo sapiens*.

On the other hand, we want to be hopeful, and hope is one of our duties. A part of our obligation to our own being and to our descendants is to study our life and our condition, searching always for the authentic underpinnings of hope. And if we look, these underpinnings can still be found.

For one thing, though we have caused the earth to be seriously diseased, it is not yet without health. The earth we have before us now is still abounding and beautiful. We must learn again to see that present world for what it is. The health of nature is the primary ground of hope—if we can find the humility and wisdom to accept nature as our teacher. The pattern of land stewardship is set by nature. This is why we must have stable rural economies and communities; we must keep alive in every place the human knowledge of the nature of that place. Nature is the best farmer and forester, for she does not destroy the land in order to make it productive. And so in our wish to preserve our land, we are not without the necessary lessons, nor are we without instruction, in our cultural and religious tradition, necessary to learn those lessons.

But we have not only the example of nature; we have still, though few and widely scattered, sufficient examples of competent and loving human stewardship of the earth. We have, too, our own desire to be healthy in a healthy world. Surely, most of us still have, somewhere within us, the fundamental human wish to die in a world in which we have been glad to live. And we *are*, in spite of much evidence to the contrary, somewhat sapient. We *can* think— if we will. If we know carefully enough who, what, and where we are, and if we keep the scale of our work small enough, we can think responsibly.

These assets are not the gigantic, technical, and costly equipment that we tend to think we need, but they are enough. They are, in fact, God's plenty. Because we have these assets, which are the supports of our legitimate hope, we can start from where we are, with what we have, and imagine and work for the healings that are necessary.

But we must begin by giving up any idea that we can bring about these healings without fundamental changes in the way we think and live. We face a choice that is starkly simple: we must change or be changed. If we fail to change for the better, then we will be changed for the worse. We cannot blunder our way into health by the same sad and foolish hopes by which we have blundered into disease. We must see that the standardless aims of industrial communism and industrial capitalism equally have failed. The aims of productivity, profitability, efficiency, limitless growth, limitless wealth, limitless power, limitless mechanization and automation can enrich and empower the few (for a while), but they will sooner or later ruin us all. The gross national product and the corporate bottom line are utterly meaningless as measures of the prosperity or health of the country.

If we want to succeed in our dearest aims and hopes as a people, we must understand that we cannot proceed any further without standards, and we must see that ultimately the standards are not set by us but by nature. We must see that it is foolish, sinful, and suicidal to destroy the health of nature for the sake of an economy that is really not an economy at all but merely a financial system, one that is unnatural, undemocratic, sacrilegious, and ephemeral. We must see the error of our effort to live by fire, by burning the world in order to live in it. There is no plainer symptom of our insanity than our avowed intention to maintain by fire an unlimited economic growth. Fire destroys what nourishes it and so in fact imposes severe limits on any growth associated with it. The true source and analogue of our economic life is the economy of plants, which never exceeds natural limits, never grows beyond the power of its place to support it, produces no waste, and enriches and preserves itself by death and decay. We must learn to grow like a tree, not like a fire. We must repudiate what Edward Abbey called "the ideology of the cancer cell": the idiotic ideology of "unlimited economic growth" that pushes blindly toward the limitation of massive catastrophe.

We must give up also our superstitious conviction that we can contrive technological solutions to all our problems. Soil loss, for example, is a problem that embarrasses all of our technological pretensions. If soil were all being lost in a huge slab somewhere, that would appeal to the would-be heroes of "science and technology," who might conceivably engineer a glamorous, large, and speedy solution — however many new problems they might cause in doing so. But soil is not usually lost in slabs or heaps of magnificent tonnage. It is lost a little at a time over millions of acres by the careless acts of millions of people. It cannot be saved by heroic feats of gigantic technology but only by millions of small acts and restraints, conditioned by small fidelities, skills, and desires. Soil loss is ultimately a cultural problem; it will be corrected only by cultural solutions.

The aims of production, profit, efficiency, economic growth, and technological progress imply, as I have said, no social or ecological standards, and in practice they submit to none. But there is another set of aims that does imply a standard, and these aims are freedom (which is pretty much a synonym for personal and local self-sufficiency), pleasure (that is, our gladness to be alive), and longevity or sustainability (by which we signify our wish that human freedom and pleasure may last). The standard implied by all of these aims is health. They depend ultimately and inescapably on the health of nature; the idea that freedom and pleasure can last long in a diseased world is preposterous. But these good things depend also on the health of human culture, and human culture is to a considerable extent the knowledge of economic and other domestic procedures — that is, ways of work, pleasure, and education — that preserve the health of nature.

In talking about health, we have thus begun to talk about community. But we must take care to see how this standard of health enlarges and clarifies the idea of community. If we speak of a *healthy* community, we cannot be speaking of a community that is merely human. We are talking about a neighborhood of humans in a place, plus the place itself: its soil, its water, its air, and all the families and tribes of the nonhuman creatures that belong to it. If the place is well preserved, if its entire membership, natural and human, is present in it, and if the human economy is in practical harmony with the nature of the place, then the community is healthy. A diseased community will be suffering natural losses that become, in turn, human losses. A healthy community is sustainable; it is, within reasonable limits, self-sufficient and, within reasonable limits, self-determined — that is, free of tyranny.

Community, then, is an indispensable term in any discussion of the connection between people and land. A healthy community is a form that includes all the local things that are connected by the larger, ultimately mysterious form of the Creation. In speaking of community, then, we are speaking of a complex connection not only among human beings or between humans and their homeland but also between the human economy and nature, between forest or prairie and field or orchard, and between troublesome creatures and pleasant ones. *All* neighbors are included.

From the standpoint of such a community, any form of land abuse—a clear-cut, a strip mine, an overplowed or overgrazed field—is as alien and as threatening as it would be from the standpoint of an ecosystem. From such a standpoint, it would be plain that land abuse reduces the possibilities of local life, just as do chain stores, absentee owners, and consolidated schools.

One obvious advantage of such an idea of community is that it provides a common ground and a common goal between conservationists and small-scale land users. The long-standing division between conservationists and farmers, ranchers, and other private small-business people is distressing because it is to a considerable extent false. It is readily apparent that the economic forces that threaten the health of ecosystems and the survival of species are equally threatening to economic democracy and the survival of human neighborhoods.

I believe that the most necessary question now—for conservationists, for small-scale farmers, ranchers, and businesspeople, for politicians interested in the survival of democracy, and for consumers—is this: What must be the economy of a healthy community based in agriculture or forestry? It *cannot* be the present colonial economy in which only "raw materials" are exported and *all* necessities and pleasures are imported. To be healthy, land-based communities will need to add value to local products, they will need to supply local demand, and they will need to be reasonably self-sufficient in food, energy, pleasure, and other basic requirements.

Once a person understands the necessity of healthy local communities and community economies, it becomes easy to imagine a range of reforms that might bring them into being.

It is at least conceivable that useful changes might be started or helped along by consumer demand in the cities. There is, for example, already evidence of a growing concern among urban consumers about the quality and the purity of food. Once this demand grows extensive and competent enough, it will have the power to change agriculture—if there is enough left of agriculture, by then, to be changed.

It is even conceivable that our people in Washington might make decisions tending toward sustainability and self-sufficiency in local economies. The federal government could do much to help, if it would. Its mere acknowledgment that problems exist would be a promising start.

But let us admit that urban consumers are not going to be well informed about their economic sources very soon and that a federal administration enlightened about the needs and problems of the countryside is not an immediate prospect.

The real improvements then must come, to a considerable extent, from the local communities themselves. We need local revision of our methods of land use and production. We need to study and work together to reduce scale, reduce overhead, reduce industrial dependencies; we need to market and process local products locally; we need to bring local economies into harmony with local ecosystems so that we can live and work with pleasure in the same places indefinitely; we need to substitute ourselves, our neighborhoods, our local resources, for expensive imported goods and services; we need to increase cooperation among all local economic entities: households, farms, factories, banks, consumers, and suppliers. If we are serious about reducing government and the burdens of government, then we need to do so by returning economic self-determination to the people. And we must not do this by inviting destructive industries to provide "jobs" in the community; we must do it by fostering economic democracy. For example, as much as possible of the food that is consumed locally ought to be locally produced on small farms, and then processed in small, nonpolluting plants that are locally owned. We must do everything possible to provide to ordinary citizens the opportunity to own a small, usable share of the country. In that way, we will put local capital to work locally, not to exploit and destroy the land but to use it well. This is not work just for the privileged, the well-positioned, the wealthy, and the powerful. It is work for everybody.

I acknowledge that to advocate such reforms is to advocate a kind of secession — not a secession of armed violence but a quiet secession by which people find the practical means and the strength of spirit to remove themselves from an economy that is exploiting them and destroying their homeland. The great, greedy, indifferent national and international economy is killing rural America, just as it is killing America's cities — it is killing our country. Experience has shown that there is no use in appealing to this economy for mercy toward the earth or toward any human community. All true patriots must find ways of opposing it.

1991

PART IV

Agrarian Economics

Much of Berry's recent work has focused on the critique of the prevailing economic order. Economics, broadly conceived as the art of household management, is the arena in which theory and practice best come together, and so represents the ideal venue for the articulation of an agrarian ethos. Berry insists that an authentic or responsible economic life requires due acknowledgment of the greater natural economy upon which all humanly derived economies depend. When this dependence is denied or ignored, as so much contemporary practice does, both economies are bound to suffer. Successful and healthy economic life, — that is, economic life that preserves the sources of life, and thus can be sustained over the long term — demands the sort of accountability and care that can only come from an agrarian focus on the local and the particular.

Economy and Pleasure

To those who still uphold the traditions of religious and political thought that influenced the shaping of our society and the founding of our government, it is astonishing, and of course discouraging, to see economics now elevated to the position of ultimate justifier and explainer of all the affairs of our daily life, and competition enshrined as the sovereign principle and ideal of economics.

As thousands of small farms and small local businesses of all kinds falter and fail under the effects of adverse economic policies or live under the threat of what we complacently call "scientific progress," the economist sits in the calm of professorial tenure and government subsidy, commenting and explaining for the illumination of the press and the general public. If those who fail happen to be fellow humans, neighbors, children of God, and citizens of the republic, all that is outside the purview of the economist. As the farmers go under, as communities lose their economic supports, as all of rural America sits as if condemned in the shadow of the "free market" and "revolutionary science," the economist announces pontifically to the press that "there will be some winners and some losers"— as if that might justify and clarify everything, or anything. The sciences, one gathers, mindlessly serve economics, and the humanities defer abjectly to the sciences. All assume, apparently, that we are in the grip of the determination of economic laws that are the laws of the universe. The newspapers quote the economists as the ultimate authorities. We read their pronouncements, knowing that the last word has been said.

"Science," President Reagan says, "tells us that the breakthroughs in superconductivity bring us to the threshold of a new age." He is speaking to "a federal conference on the commercial applications of the new technology," and we know that by "science" he means scientists in the pay of corporations. "It is our task at this conference," he says, "to herald in that new age with a rush." A part of his program to accomplish this task is a proposal to "relax" the antitrust laws.* Thus even the national executive and our legal system itself must now defer to the demands of "the economy." Whatever "new age" is at hand at the moment must be heralded in "with a rush" because of the profits available to those who will rush it in.

It seems that we have been reduced almost to a state of absolute economics, in which people and all other creatures and things may be considered purely as economic "units," or integers of production, and in which a human being may be dealt with, as John Ruskin put it, "merely as a covetous machine."† And the voices bitterest to hear are those saying that all this destructive work of mindless genius, money, and power is regrettable but cannot be helped.

Perhaps it cannot. Surely we would be fools if, having understood the logic of this terrible process, we assumed that it might not go on in its glutton's optimism until it achieves the catastrophe that is its logical end. But let us suppose that a remedy is possible. If so, perhaps the best beginning would be in understanding the falseness and silliness of the economic ideal of competition, which is destructive both of nature and of human nature because it is untrue to both.

The ideal of competition always implies, and in fact requires, that any community must be divided into a class of winners and a class of losers. This division is radically different from other social divisions: that of the more able and the less able, or that of the richer and the poorer, or even that of the rulers and the ruled. These latter divisions have existed throughout history and at times, at least, have been ameliorated by social and religious ideals that instructed the strong to help the weak. As a purely economic ideal, competition does not contain or imply any such instructions. In fact, the defenders of the ideal of

*"Reagan calls for effort to find commercial uses for superconductors," *Louisville Courier-Journal*, July 29, 1987, p. A3.

†John Ruskin, *Unto This Last* (Lincoln: University of Nebraska Press, 1967), p. 11.

competition have never known what to do with or for the losers. The losers simply accumulate in human dumps, like stores of industrial waste, until they gain enough misery and strength to overpower the winners. The idea that the displaced and dispossessed "should seek retraining and get into another line of work" is, of course, utterly cynical; it is only the hand-washing practiced by officials and experts.* A loser, by definition, is somebody whom nobody knows what to do with. There is no limit to the damage and the suffering implicit in this willingness that losers should exist as a normal economic cost.

The danger of the ideal of competition is that it neither proposes nor implies any limits. It proposes simply to lower costs at any cost, and to raise profits at any cost. It does not hesitate at the destruction of the life of a family or the life of a community. It pits neighbor against neighbor as readily as it pits buyer against seller. Every transaction is *meant* to involve a winner and a loser. And for this reason the human economy is pitted without limit against nature. For in the unlimited competition of neighbor and neighbor, buyer and seller, all available means must be used; none may be spared.

I will be told that indeed there are limits to economic competitiveness as now practiced—that, for instance, one is not allowed to kill one's competitor. But, leaving aside the issue of whether or not murder would be acceptable as an economic means if the stakes were high enough, it is a fact that the destruction of life is a part of the daily business of economic competition as now practiced. If one person is willing to take another's property or to accept another's ruin as a normal result of economic enterprise, then he is willing to destroy that other person's life as it is and as it desires to be. That this person's biological existence has been spared seems merely incidental; it was spared because it was not worth anything. That this person is now "free" to "seek retraining and get into another line of work" signifies only that his life as it was has been destroyed.

But there is another implication in the limitlessness of the ideal of competition that is politically even more ominous: namely, that unlimited economic competitiveness proposes an unlimited concentration of economic power. Economic anarchy, like any other free-for-all, tends inevitably toward dominance by the strongest. If it is normal for economic activity to divide the community into a class of winners and a class of losers, then the inescapable implication is that the class of winners will become ever smaller, the class of losers ever larger. And that, obviously, is now happening: the usable property of our

*Reed Karaim, "Loss of million farms in 14 years projected," *Des Moines Register*, March 18, 1986, p. 1A.

country, once divided somewhat democratically, is owned by fewer and fewer people every year. That the president of the republic can, without fear, propose the "relaxation" of antitrust laws in order to "rush" the advent of a commercial "new age" suggests not merely that we are "rushing" toward plutocracy, but that this is now a permissible goal for the would-be winning class for which Mr. Reagan speaks and acts, and a burden acceptable to nearly everybody else.

Nowhere, I believe, has this grossly oversimplified version of economics made itself more at home than in the land-grant universities. The colleges of agriculture, for example, having presided over the now nearly completed destruction of their constituency—the farm people and the farm communities—are now scrambling to ally themselves more firmly than ever, not with "the rural home and rural life"* that were, and are, their trust, but with the technocratic aims and corporate interests that are destroying the rural home and rural life. This, of course, is only a new intensification of an old alliance. The revolution that began with machines and chemicals proposes now to continue with automation, computers, and biotechnology. That this has been and is a revolution is undeniable. It has not been merely a "scientific revolution," as its proponents sometimes like to call it, but also an economic one, involving great and profound changes in property ownership and the distribution of real wealth. It has done by insidious tendency what the communist revolutions have done by fiat: it has dispossessed the people and usurped the power and integrity of community life.

This work has been done, and is still being done, under the heading of altruism: its aims, as its proponents never tire of repeating, are to "serve agriculture" and to "feed the world." These aims, as stated, are irreproachable; as pursued, they raise a number of doubts. Agriculture, it turns out, is to be served strictly according to the rules of competitive economics. The aim is "to make farmers more competitive" and "to make American agriculture more competitive." Against whom, we must ask, are our farmers and our agriculture to be made more competitive? And we must answer, because we know: Against other farmers, at home and abroad. Now, if the colleges of agriculture "serve agriculture" by helping farmers to compete against one another, what do they propose to do to help the farmers who have been out-competed? Well, those people are not farmers anymore, and therefore are of no concern to the academic servants of agriculture. Besides, they are the beneficiaries of the inestimable liberty to "seek retraining and get into another line of work."

*This is the language of the Hatch Act, *United States Code*, Section 361b.

And so the colleges of agriculture, entrusted though they are to serve the rural home and rural life, give themselves over to a hysterical rhetoric of "change," "the future," "the frontiers of modern science," "competition," "the competitive edge," "the cutting edge," "early adoption," and the like, as if there is nothing worth learning from the past and nothing worth preserving in the present. The idea of the teacher and scholar as one called upon to preserve and pass on a common cultural and natural birthright has been almost entirely replaced by the idea of the teacher and scholar as a developer of "human capital" and a bestower of economic advantage. The ambition is to make the university an "economic resource" in a competition for wealth and power that is local, national, and global. Of course, all this works directly against the rural home and rural life, because it works directly against community.

There is no denying that competitiveness is a part of the life both of an individual and of a community, or that, within limits, it is a useful and necessary part. But it is equally obvious that no individual can lead a good or a satisfying life under the rule of competition, and that no community can succeed except by limiting somehow the competitiveness of its members. One cannot maintain one's "competitive edge" if one helps other people. The advantage of "early adoption" would disappear — it would not be thought of — in a community that put a proper value on mutual help. Such advantages would not be thought of by people intent on loving their neighbors as themselves. And it is impossible to imagine that there can be any reconciliation between local and national competitiveness and global altruism. The ambition to "feed the world" or "feed the hungry," rising as it does out of the death struggle of farmer with farmer, proposes not the filling of stomachs, but the engorgement of "the bottom line." The strangest of all the doctrines of the cult of competition, in which admittedly there must be losers as well as winners, is that the result of competition is inevitably good for everybody, that altruistic ends may be met by a system without altruistic motives or altruistic means.

In agriculture, competitiveness has been based throughout the industrial era on constantly accelerating technological change — the very *principle* of agricultural competitiveness is ever-accelerating change — and this has encouraged an ever-accelerating dependency on purchased products, products purchased ever farther from home. Community, however, aspires toward stability. It strives to balance change with constancy. That is why community life places such high value on neighborly love, marital fidelity, local loyalty, the integrity and continuity of family life, respect for the old, and instruction of the young. And a vital community draws its life, so far as possible, from local sources. It prefers to solve its problems, for example, by nonmonetary exchanges of help, not by buying things. A community cannot survive under the rule of competition.

But the land-grant universities, in espousing the economic determinism of the industrialists, have caught themselves in a logical absurdity that they may finally discover to be dangerous to themselves. If competitiveness is the economic norm, and the "competitive edge" the only recognized social goal, then how can these institutions justify public support? Why, in other words, should the public be willing to permit a corporation to profit privately from research that has been subsidized publicly? Why should not the industries be required to afford their own research, and why should not the laws of competition and the free market—if indeed they perform as advertised—enable industries to do their own research a great deal more cheaply than the universities can do it?

* * *

The question that we finally come to is a practical one, though it is not one that is entirely answerable by empirical methods: Can a university, or a nation, *afford* this exclusive rule of competition, this purely economic economy? The great fault of this approach to things is that it is so drastically reductive; it does not permit us to live and work as human beings, as the best of our inheritance defines us. Rats and roaches live by competition under the law of supply and demand; it is the privilege of human beings to live under the laws of justice and mercy. It is impossible not to notice how little the proponents of the ideal of competition have to say about honesty, which is the fundamental economic virtue, and how *very* little they have to say about community, compassion, and mutual help.

But what the ideal of competition most flagrantly and disastrously excludes is affection. The affections, John Ruskin said, are "an anomalous force, rendering every one of the ordinary political economist's calculations nugatory; while, even if he desired to introduce this new element into his estimates, he has no power of dealing with it; for the affections only become a true motive power when they ignore every other motive power and condition of political economy." Thus, if we are sane, we do not dismiss or abandon our infant children or our aged parents because they are too young or too old to work. For human beings, affection is the ultimate motive, because the force that powers us, as Ruskin also said, is not "steam, magnetism, or gravitation," but "a Soul."

I would like now to attempt to talk about economy from the standpoint of affection—or, as I am going to call it, pleasure, advancing just a little beyond

Ruskin's term, for pleasure is, so to speak, affection in action. There are obvious risks in approaching an economic problem by a way that is frankly emotional—to talk, for example, about the pleasures of nature and the pleasures of work. But these risks seem to me worth taking, for what I am trying to deal with here is the grief that we increasingly suffer as a result of the loss of those pleasures.

It is necessary, at the outset, to make a distinction between pleasure that is true or legitimate and pleasure that is not. We know that a pleasure can be as heavily debited as an economy. Some people undoubtedly thought it pleasant, for example, to have the most onerous tasks of their economy performed by black slaves. But this proved to be a pleasure that was temporary and dangerous. It lived by an enormous indebtedness that was inescapably to be paid not in money, but in misery, waste, and death. The pleasures of fossil fuel combustion and nuclear "security" are, as we are beginning to see, similarly debited to the future. These pleasures are in every way analogous to the self-indulgent pleasures of individuals. They are pleasures that we are allowed to have merely to the extent that we can ignore or defer the logical consequences.

That there is pleasure in competition is not to be doubted. We know from childhood that winning is fun. But we probably begin to grow up when we begin to sympathize with the loser—that is, when we begin to understand that competition involves costs as well as benefits. Sometimes perhaps, as in the most innocent games, the benefits are all to the winner and the costs all to the loser. But when the competition is more serious, when the stakes are higher and greater power is used, then we know that the winner shares in the cost, sometimes disastrously. In war, for example, even the winner is a loser. And this is equally true of our present economy: in unlimited economic competition, the winners are losers; that they may appear to be winners is owing only to their temporary ability to charge their costs to other people or to nature.

But a victory over community or nature can be won only at everybody's cost. For example, we now have in the United States many landscapes that have been defeated—temporarily or permanently—by strip mining, by clear-cutting, by poisoning, by bad farming, or by various styles of "development" that have subjugated their sites entirely to human purposes. These landscapes have been defeated for the benefit of what are assumed to be victorious landscapes: the suburban housing developments and the places of amusement (the park systems, the recreational wildernesses) of the winners—so far—in the economy. But these victorious landscapes and their human inhabitants are already paying the costs of their defeat of other landscapes: in air and water pollution, overcrowding, inflated prices, and various diseases of body and mind. Eventually, the cost will be paid in scarcity or want of necessary goods.

Is it possible to look beyond this all-consuming "rush" of winning and losing to the possibility of countrysides, a nation of countrysides, in which use is not synonymous with defeat? It is. But in order to do so we must consider our pleasures. Since we all know, from our own and our nation's experience, of some pleasures that are canceled by their costs, and of some that result in unredeemable losses and miseries, it is natural to wonder if there may not be such phenomena as *net* pleasures, pleasures that are free or without a permanent cost. And we know that there are. These are the pleasures that we take in our own lives, our own wakefulness in this world, and in the company of other people and other creatures—pleasures innate in the Creation and in our own good work. It is in these pleasures that we possess the likeness to God that is spoken of in Genesis.

"This curious world we inhabit is more wonderful than convenient; more beautiful than it is useful; it is more to be admired and enjoyed than used." Henry David Thoreau said that to his graduating class at Harvard in 1837. We may assume that to most of them it sounded odd, as to most of the Harvard graduating class of 1987 it undoubtedly still would. But perhaps we will be encouraged to take him seriously, if we recognize that this idea is not something that Thoreau made up out of thin air. When he uttered it, he may very well have been remembering Revelation 4:11: "Thou art worthy, O Lord, to receive glory and honour and power: for thou hast created all things, and for thy pleasure they are and were created." That God created "all things" is in itself an uncomfortable thought, for in our workaday world we can hardly avoid preferring some things above others, and this makes it hard to imagine *not* doing so. That God created all things for His pleasure, and that they continue to exist because they please Him, is formidable doctrine indeed, as far as possible both from the "anthropocentric" utilitarianism that some environmentalist critics claim to find in the Bible and from the grouchy spirituality of many Christians.

It would be foolish, probably, to suggest that God's pleasure in all things can be fully understood or appreciated by mere humans. The passage suggests, however, that our truest and profoundest religious experience may be the simple, unasking pleasure in the existence of other creatures that *is* possible to humans. It suggests that God's pleasure in all things must be respected by us in our use of things, and even in our displeasure in some things. It suggests too that we have an obligation to preserve God's pleasure in all things, and surely this means not only that we must not misuse or abuse anything, but also that there must be some things and some places that by common agreement we do not use at all, but leave wild. This bountiful and lovely thought that all creatures are pleasing to God—and potentially pleasing, therefore, to us—is unthinkable from the point of view of an economy divorced from pleasure, such as the one we have now, which completely discounts the capacity of people to be affectionate toward what they do and what they use and where they live and the other people and creatures with whom they live.

It may be argued that our whole society is more devoted to pleasure than any whole society ever was in the past, that we support in fact a great variety of pleasure industries and that these are thriving as never before. But that would seem only to prove my point. That there can be pleasure industries at all, exploiting our apparently limitless inability to be pleased, can only mean that our economy is divorced from pleasure and that pleasure is gone from our workplaces and our dwelling places. Our workplaces are more and more exclusively given over to production, and our dwelling places to consumption. And this accounts for the accelerating division of our country into defeated landscapes and victorious (but threatened) landscapes.

More and more, we take for granted that work must be destitute of pleasure. More and more, we assume that if we want to be pleased we must wait until evening, or the weekend, or vacation, or retirement. More and more, our farms and forests resemble our factories and offices, which in turn more and more resemble prisons—why else should we be so eager to escape them? We recognize defeated landscapes by the absence of pleasure from them. We are defeated at work because our work gives us no pleasure. We are defeated at home because we have no pleasant work there. We turn to the pleasure industries for relief from our defeat, and are again defeated, for the pleasure industries can thrive and grow only upon our dissatisfaction with them.

Where is our comfort but in the free, uninvolved, finally mysterious beauty and grace of this world that we did not make, that has no price? Where is our sanity but there? Where is our pleasure but in working and resting kindly in the presence of this world?

And in the right sort of economy, our pleasure would not be merely an addition or by-product or reward; it would be both an empowerment of our work and its indispensable measure. Pleasure, Ananda Coomaraswamy said, *perfects* work. In order to have leisure and pleasure, we have mechanized and automated and computerized our work. But what does this do but divide us ever more from our work and our products—and, in the process, from one another and the world? What have farmers done when they have mechanized and computerized their farms? They have removed themselves and their pleasure from their work.

I was fortunate, late in his life, to know Henry Besuden of Clark County, Kentucky, the premier Southdown sheep breeder and one of the great farmers of his time. He told me once that his first morning duty in the spring and early summer was to saddle his horse and ride across his pastures to see the condition of the grass when it was freshest from the moisture and coolness of the night. What he wanted to see in his pastures at that time of year, when his spring lambs would be fattening, was what he called "bloom"—by which he meant not flowers, but a certain visible delectability. He recognized it, of course, by his delight in it. He was one of the best of the traditional livestockmen—the husbander or husband of his animals. As such, he was not interested in "statistical indicators" of his flock's "productivity." He wanted his sheep to be pleased. If they were pleased with their pasture, they would eat eagerly, drink well, rest, and grow. He knew their pleasure by his own.

The nearly intolerable irony in our dissatisfaction is that we have removed pleasure from our work in order to remove "drudgery" from our lives. If I could pick any rule of industrial economics to receive a thorough reexamination by our people, it would be the one that says that all hard physical work is "drudgery" and not worth doing. There are of course many questions surrounding this issue: What is the work? In whose interest is it done? Where and in what circumstances is it done? How well and to what result is it done? In whose company is it done? How long does it last? And so forth. But this issue is personal and so needs to be reexamined by everybody. The argument, if it is that, can proceed only by personal testimony.

I can say, for example, that the tobacco harvest in my own home country involves the hardest work that I have done in any quantity. In most of the years of my life, from early boyhood until now, I have taken part in the tobacco cutting. This work usually occurs at some time between the last part of August and the first part of October. Usually the weather is hot; usually we are in a hurry. The work is extremely demanding, and often, because of the weather, it has the character of an emergency. Because all of the work still must be done by hand, this event has maintained much of its old character; it is very much the sort of thing the agriculture experts have had in mind when they have talked about freeing people from drudgery.

That the tobacco cutting *can* be drudgery is obvious. If there is too much of it, if it goes on too long, if one has no interest in it, if one cannot reconcile oneself to the misery involved in it, if one does not like or enjoy the company of one's fellow workers, then drudgery would be the proper name for it.

But for me, and I think for most of the men and women who have been my companions in this work, it has not been drudgery. None of us would say that we take pleasure in all of it all of the time, but we do take pleasure in it, and sometimes the pleasure can be intense and clear. Many of my dearest memories come from these times of hardest work.

The tobacco cutting is the most protracted social occasion of our year. Neighbors work together; they are together all day every day for weeks. The quiet of the work is not much interrupted by machine noises, and so there is much talk. There is the talk involved in the management of the work. There is incessant speculation about the weather. There is much laughter; because of the unrelenting difficulty of the work, everything funny or amusing is relished. And there are memories.

The crew to which I belong is the product of kinships and friendships going far back; my own earliest associations with it occurred nearly forty years ago. And so as we work we have before us not only the present crop and the present fields, but other crops and other fields that are remembered. The tobacco cutting is a sort of ritual of remembrance. Old stories are retold; the dead and the absent are remembered. Some of the best talk I have ever listened to I have heard during these times, and I am especially moved to think of the care that is sometimes taken to speak well—that is, to speak fittingly—of the dead and the absent. The conversation, one feels, is ancient. Such talk in barns and at row ends must go back without interruption to the first farmers. How long it may continue is now an uneasy question; not much longer perhaps, but we do not know. We only know that while it lasts it can carry us deeply into our shared life and the happiness of farming.

On many days we have had somebody's child or somebody's children with us, playing in the barn or around the patch while we worked, and these have been our best days. One of the most regrettable things about the industrialization of work is the segregation of children. As industrial work excludes the dead by social mobility and technological change, it excludes children by haste and danger. The small scale and the handwork of our tobacco cutting permit margins both temporal and spatial that accommodate the play of children. The children play at the grown-ups' work, as well as at their own play. In their play the children learn to work; they learn to know their elders and their country. And the presence of playing children means invariably that the grown-ups play too from time to time.

(I am perforce aware of the problems and the controversies about tobacco. I have spoken of the tobacco harvest here simply because it is the only remaining farm job in my part of the country that still involves a traditional neighborliness.)

Ultimately, in the argument about work and how it should be done, one has only one's pleasure to offer. It is possible, as I have learned again and again, to be in one's place, in such company, wild or domestic, and with such pleasure, that one cannot think of another place that one would prefer to be—or of another place at all. One does not miss or regret the past, or fear or long for the future. Being there is simply all, and is enough. Such times give one the chief standard and the chief reason for one's work.

Last December, when my granddaughter, Katie, had just turned five, she stayed with me one day while the rest of the family was away from home. In the afternoon we hitched a team of horses to the wagon and hauled a load of dirt for the barn floor. It was a cold day, but the sun was shining; we hauled our load of dirt over the tree-lined gravel lane beside the creek—a way well known to her mother and to my mother when they were children. As we went along, Katie drove the team for the first time in her life. She did very well, and she was proud of herself. She said that her mother would be proud of her, and I said that I was proud of her.

We completed our trip to the barn, unloaded our load of dirt, smoothed it over the barn floor, and wetted it down. By the time we started back up the creek road the sun had gone over the hill and the air had turned bitter. Katie sat close to me in the wagon, and we did not say anything for a long time. I did not say anything because I was afraid that Katie was not saying anything because she was cold and tired and miserable and perhaps homesick; it was impossible to hurry much, and I was unsure how I would comfort her.

But then, after a while, she said, "Wendell, isn't it fun?"

1988

Two Economies (1983)

I

Some time ago, in a conversation with Wes Jackson in which we were laboring to define the causes of the modern ruination of farmland, we finally got around to the money economy. I said that an economy based on energy would be more benign because it would be more comprehensive.

Wes would not agree. "An energy economy still wouldn't be comprehensive enough."

"Well," I said, "then what kind of economy *would* be comprehensive enough?"

He hesitated a moment, and then, grinning, said, "The Kingdom of God."

I assume that Wes used the term because he found it, at that point in our conversation, indispensable; I assume so because, in my pondering over its occurrence at that point, I have found it indispensable myself. For the thing that troubles us about the industrial economy is exactly that it is not comprehensive enough, that, moreover, it tends to destroy what it does not comprehend, and that it is *dependent* upon much that it does not comprehend. In attempting to criticize such an economy, we naturally pose against it an economy that does not leave anything out, and we can say without presuming too much that the first principle of the Kingdom of God is that it includes everything; in it, the fall of every sparrow is a significant event. We are in it whether we know it or not and whether we wish to be or not. Another principle, both ecological and traditional, is that everything in the Kingdom of God is joined both to it and to everything else that is in it; that is to say, the Kingdom of God is orderly. A third principle is that humans do not and can never know either all the creatures that the Kingdom of God contains or the whole pattern or order by which it contains them.

The suitability of the Kingdom of God as, so to speak, a place name is partly owing to the fact that it still means pretty much what it has always meant. Because, I think, of the embarrassment that the phrase has increasingly caused among the educated, it has not been much tainted or tampered with by the disinterested processes of academic thought; it is a phrase that comes to us with its cultural strings still attached. To say that we live in the Kingdom of God is both to suggest the difficulty of our condition and to imply a fairly complete set of culture-borne instructions for living in it. These instructions are not always explicitly ecological, but it can be argued that they are always implicitly so, for all of them rest ultimately on the assumptions that I have given as the second and third principles of the Kingdom of God: that we live within order and that this order is both greater and more intricate than we can know. The difficulty of our predicament, then, is made clear if we add a fourth principle: though we cannot produce a complete or even an adequate description of this order, severe penalties are in store for us if we presume upon it or violate it.

I am not dealing, of course, with perceptions that are only Biblical. The ancient Greeks, according to Aubrey de Sélincourt, saw "a continuing moral pattern in the vicissitudes of human fortune," a pattern "formed from the belief that men, as men, are subject to certain limitations imposed by a Power —call it Fate or God—which they cannot fully comprehend, and that any attempt to transcend those limitations is met by inevitable punishment." The Greek name for the pride that attempts to transcend human limitations was *hubris*, and hubris was the cause of what the Greeks understood as tragedy.

Nearly the same sense of *necessary* human limitation is implied in the Old Testament's repeated remonstrances against too great a human confidence in the power of "mine own hand." Gideon's army against the Midianites, for example, was reduced from thirty-two thousand to three hundred expressly to prevent the Israelites from saying, "Mine own hand hath saved me." A similar purpose was served by the institution of the Sabbath, when, by not working, the Israelites were meant to see the limited efficacy of their work and thus to understand their true dependence.

* * *

Though I hope that my insistence on the usefulness of the term *the Kingdom of God* will be understood, I must acknowledge that the term is local, in the sense that it is fully available only to those whose languages are involved in Western or Biblical tradition. A person of Eastern heritage, for example, might speak of the totality of all creation, visible and invisible, as "the Tao." I am well aware also that many people would not willingly use either term, or any such term. For these reasons, I do not want to make a statement that is specially or exclusively Biblical, and so I would like now to introduce a more culturally neutral term for that economy that I have been calling the Kingdom of God. Sometimes, in thinking about it, I have called it the Great Economy, which is the name I am going to make do with here—though I will remain under the personal necessity of Biblical reference. And that, I think, must be one of my points: we can name it whatever we wish, but we cannot define it except by way of a religious tradition. The Great Economy, like the Tao or the Kingdom of God, is both known and unknown, visible and invisible, comprehensible and mysterious. It is, thus, the ultimate condition of our experience and of the practical questions rising from our experience, and it imposes on our consideration of those questions an extremity of seriousness and an extremity of humility.

I am assuming that the Great Economy, whatever we may name it, is indeed—and in ways that are, to some extent, practical—an economy: it includes principles and patterns by which values or powers or necessities are parceled out and exchanged. But if the Great Economy comprehends humans and thus cannot be fully comprehended by them, then it is also not an economy in which humans can participate directly. What this suggests, in fact, is that humans can live in the Great Economy only with great uneasiness, subject to powers and laws that they can understand only in part. There is no human accounting for the Great Economy. This obviously is a description of the circumstance of religion, the circumstance that *causes* religion. De Sélincourt states the problem succinctly: "Religion in every age is concerned with the vast and fluctuant regions of experience which knowledge cannot penetrate, the regions which a man knows, or feels, to stretch away beyond the narrow, closed circle of what he can *manage* by the use of his wits."

If there is no denying our dependence on the Great Economy, there is also no denying our need for a little economy—a narrow circle within which things are manageable by the use of our wits. I don't think Wes Jackson was denying this need when he invoked the Kingdom of God as the complete economy; rather, he was, I think, insisting upon a priority that is both proper and practical. If he had a text in mind, it must have been the sixth chapter of Matthew, in which, after speaking of God's care for nature, the fowls of the air and the lilies of the field, Jesus says: "Therefore take no thought, saying, What shall we eat? or, What shall we drink? or, Wherewithal shall we be clothed? . . . But seek ye first the kingdom of God, and his righteousness; and all these things shall be added unto you."

There is an attitude that sees in this text a denial of the value of *any* economy of this world, but this attitude makes the text useless and meaningless to humans who must live in this world. These verses make usable sense only if we read them as a statement of considerable practical import about the real nature of worldly economy. If this passage meant for us to seek *only* the Kingdom of God, it would have the odd result of making good people not only feckless but also dependent upon bad people busy with quite other seekings. It says, rather, to seek the Kingdom of God *first;* that is, it gives an obviously necessary priority to the Great Economy over any little economy made within it. The passage also clearly includes nature within the Great Economy, and it affirms the goodness, indeed the sanctity, of natural creatures.

The fowls of the air and the lilies of the field live within the Great Economy entirely by nature, whereas humans, though entirely dependent upon it, must live in it partly by artifice. The birds can live in the Great Economy only as birds, the flowers only as flowers, the humans only as humans. The humans, unlike the wild creatures, may choose not to live in it—or, rather, since no creature can escape it, they may choose to *act* as if they do not, or they may choose to try to live in it on their own terms. If humans choose to live in the Great Economy on *its* terms, then they must live in harmony with it, maintaining it in trust and learning to consider the lives of the wild creatures.

Certain economic restrictions are clearly implied, and these restrictions have mainly to do with the economics of futurity. We know from other passages in the Gospels that a certain preparedness or provisioning for the future is required of us. It may be that such preparedness is part of our obligation to today, and for *that* reason we need "take no thought for the morrow." But it is clear that such preparations can be carried too far, that we can provide too much for the future. The sin of "a certain rich man" in the twelfth chapter of Luke is that he has "much goods laid up for many years" and thus believes that he can "eat, drink, and be merry." The offense seems to be that he has stored up too much and in the process has belittled the future, for he has reduced it to the size of his own hopes and expectations. He is prepared for a future in which he will be prosperous, not for one in which he will be dead. We know from our own experience that it is possible to live in the present in such a way as to diminish the future practically as well as spiritually. By laying up "much goods" in the present — and, in the process, *using up* such goods as topsoil, fossil fuel, and fossil water — we incur a debt to the future that we cannot repay. That is, we diminish the future by deeds that we call "use" but that the future will call "theft." We may say, then, that we seek the Kingdom of God, in part, by our economic behavior, and we fail to find it if that behavior is wrong.

If we read Matthew 6:24–34 as a teaching that is *both* practical and spiritual, as I think we must, then we must see it as prescribing the terms of a kind of little economy or human economy. Since I am deriving it here from a Christian text, we could call it a Christian economy. But we need not call it that. A Buddhist might look at the working principles of the economy I am talking about and call it a Buddhist economy. E. F. Schumacher, in fact, says that the aim of "Buddhist economics" is "to obtain the maximum of well-being with the minimum of consumption," which I think is partly the sense of Matthew 6:24–34. Or we could call this economy (from Matthew 6:28) a "considerate" economy or, simply, a good economy. Whatever the name, the human economy, if it is to be a good economy, must fit harmoniously within and must correspond to the Great Economy; in certain important ways, it must be an analogue of the Great Economy.

A fifth principle of the Great Economy that must now be added to the previous four is that *we* cannot foresee an end to it: the same basic stuff is going to be shifting from one form to another, so far as we know, forever. From a human point of view, this is a rather heartless endurance. As cynics sometimes point out, conservation is always working, for what is lost or wasted in one place always turns up someplace else. Thus, soil erosion in Iowa involves no loss because the soil is conserved in the Gulf of Mexico. Such people like to point out that soil erosion is as "natural" as birdsong. And so it is, though these people neglect to observe that soil conservation is also natural, and that, before the advent of farming, nature alone worked effectively to keep Iowa topsoil in Iowa. But to say that soil erosion is natural is only a way of saying that there are some things that the Great Economy cannot do for humans. Only a little economy, only a good human economy, can define for us the value of keeping the topsoil where it is.

A good human economy, that is, defines and values human goods, and, like the Great Economy, it conserves and protects its goods. It proposes to endure. Like the Great Economy, a good human economy does not propose for itself a term to be set by humans. That termlessness, with all its implied human limits and restraints, is a human good.

The difference between the Great Economy and any human economy is pretty much the difference between the goose that laid the golden egg and the golden egg. For the goose to have value as a layer of golden eggs, she must be a live goose and therefore joined to the life cycle, which means that she is joined to all manner of things, patterns, and processes that sooner or later surpass human comprehension. The golden egg, on the other hand, can be fully valued by humans according to kind, weight, and measure—but it will not hatch, and it cannot be eaten. To make the value of the egg *fully* accountable, then, we must make it "golden," must remove it from life. But if in our valuation of it we wish to consider its relation to the goose, we have to undertake a different kind of accounting, more exacting if less exact. That is, if we wish to value the egg in such a way as to preserve the goose that laid it, we find that we must behave, not scientifically, but humanely; we must understand ourselves as humans as fully as our traditional knowledge of ourselves permits. We participate in our little human economy to a considerable extent, that is, by factual knowledge, calculation, and manipulation; our participation in the Great Economy also requires those things, but requires as well humility, sympathy, forbearance, generosity, imagination.

Another critical difference, implicit in the foregoing, is that, though a human economy can evaluate, distribute, use, and preserve things of value, it cannot make value. Value can originate only in the Great Economy. It is true enough that humans can add value to natural things: we may transform trees into boards, and transform boards into chairs, adding value at each transformation. In a good human economy, these transformations would be made by good work, which would be properly valued and the workers properly rewarded. But a good human economy would recognize at the same time that it was dealing all along with materials and powers that it did not make. It did not make trees, and it did not make the intelligence and talents of the human workers. What the humans have added at every step is artificial, made by art, and though the value of art is critical to human life, it is a secondary value.

When humans presume to originate value, they make value that is first abstract and then false, tyrannical, and destructive of real value. Money value, for instance, can be said to be true only when it justly and stably represents the value of necessary goods, such as clothing, food, and shelter, which originate ultimately in the Great Economy. Humans can originate money value in the abstract, but only by inflation and usury, which falsify the value of necessary things and damage their natural and human sources. Inflation and usury and the damages that follow can be understood, perhaps, as retributions for the presumption that humans can make value.

* * *

We may say, then, that a human economy originates, manages, and distributes secondary or added values but that, if it is to last long, it must also manage in such a way as to make continuously available those values that are primary or given, the secondary values having mainly to do with husbandry and trusteeship. A little economy is obliged to receive them gratefully and to use them in such a way as not to diminish them. We might make a long list of things that we would have to describe as primary values, which come directly into the little economy from the Great, but the one I want to talk about, because it is the one with which we have the most intimate working relationship, is the topsoil.

We cannot speak of topsoil, indeed we cannot know what it is, without acknowledging at the outset that we cannot make it. We can care for it (or not), we can even, as we say, "build" it, but we can do so only by assenting to, preserving, and perhaps collaborating in its own processes. To those processes themselves we have nothing to contribute. We cannot make topsoil, and we cannot make any substitute for it; we cannot do what it does. It is apparently impossible to make an adequate description of topsoil in the sort of language that we have come to call "scientific." For, although any soil sample can be reduced to its inert quantities, a handful of the real thing has life in it; it is full of living creatures. And if we try to describe the behavior of that life we will see that it is doing something that, if we are not careful, we will call "unearthly": it is making life out of death. Not so very long ago, had we known about it what we know now, we would probably have called it "miraculous." In a time when death is looked upon with almost universal enmity, it is hard to believe that the land we live on and the lives we live are the gifts of death. Yet that is so, and it is the topsoil that makes it so. In fact, in talking about topsoil, it is hard to avoid the language of religion. When, in "This Compost," Whitman says, "The resurrection of the wheat appears with pale visage out of its graves," he is speaking in the Christian tradition, and yet he is describing what happens, with language that is entirely accurate and appropriate. And when at last he says of the earth that "It gives such divine materials to men," we feel that the propriety of the words comes not from convention but from the actuality of the uncanny transformation that his poem has required us to imagine, as if in obedience to the summons to "consider the lilies of the field."

Even in its functions that may seem, to mechanists, to be mechanical, the topsoil behaves complexly and wonderfully. A healthy topsoil, for instance, has at once the ability to hold water and to drain well. When we speak of the health of a watershed, these abilities are what we are talking about, and the word "health," which we do use in speaking of watersheds, warns us that we are not speaking merely of mechanics. A healthy soil is made by the life dying into it and by the life living in it, and to its double ability to drain and retain water we are complexly indebted, for it not only gives us good crops but also erosion control as well as *both* flood control and a constant water supply.

Obviously, topsoil, not energy or money, is the critical quantity in agriculture. And topsoil *is* a quantity; we need it in quantities. We now need more of it than we have; we need to help it to make more of itself. But it is a most peculiar quantity, for it is inseparable from quality. Topsoil is by definition *good* soil, and it can be preserved in human use only by good care. When humans see it as a mere quantity, they tend to make it that; they destroy the life in it, and they begin to measure in inches and feet and tons how much of it they have "lost."

When we see the topsoil as the foundation of that household of living creatures and their nonliving supports that we now call an "ecosystem" but which some of us understand better as a "neighborhood," we find ourselves in debt for other benefits that baffle our mechanical logic and defy our measures. For example, one of the principles of an ecosystem is that diversity increases capacity—or, to put it another way, that complications of form or pattern can increase greatly within quantitative limits. I suppose that this may be true only up to a point, but I suppose also that that point is far beyond the human capacity to understand or diagram the pattern.

On a farm put together on a sound ecological pattern, the same principle holds. Henry Besuden, the great farmer and shepherd of Clark County, Kentucky, compares the small sheep flock to the two spoons of sugar that can be added to a brimful cup of coffee, which then becomes "more palatable [but] doesn't run over. You can stock your farm to the limit with other livestock and still add a small flock of sheep." He says this, characteristically, after rejecting the efforts of sheep specialists to get beyond "the natural physical limits of the ewe" by breeding out of season in order to get three lamb crops in two years or by striving for "litters" of lambs rather than nature's optimum of twins. Rather than chafe at "natural physical limits," he would turn to nature's elegant way of enriching herself *within* her physical limits by diversification, by complication of pattern. Rather than strain the productive capacity of the ewe, he would, without strain, enlarge the productive capacity of the farm — a healthier, safer, and cheaper procedure. Like many of the better traditional farmers, Henry Besuden is suspicious of "the measure of land in length and width," for he would be mindful as well of "the depth and quality."

A small flock of ewes, fitted properly into a farm's pattern, virtually disappears into the farm and does it good, just as it virtually disappears into the time and energy economy of a farm family and does it good. And, properly fitted into the farm's pattern, the small flock virtually disappears from the debit side of the farm's accounts but shows up plainly on the credit side. This "disappearance" is possible, not to the extent that the farm is a human artifact, a belonging of the human economy, but to the extent that it remains, by its obedience to natural principle, a belonging of the Great Economy.

A little economy may be said to be good insofar as it perceives the excellence of these benefits and husbands and preserves them. It is by holding up this standard of goodness that we can best see what is wrong with the industrial economy. For the industrial economy does not see itself as a little economy; it sees itself as the *only* economy. It makes itself thus exclusive by the simple expedient of valuing only what it can use — that is, only what it can regard as "raw material" to be transformed mechanically into something else. What it cannot use, it characteristically describes as "useless," "worthless," "random," or "wild," and gives it some such name as "chaos," "disorder," or "waste"— and thus ruins it or cheapens it in preparation for eventual use. That western deserts or eastern mountains were once perceived as "useless" made it easy to dignify them by the "use" of strip mining. Once we acknowledge the existence of the Great Economy, however, we are astonished and frightened to see how much modern enterprise is the work of hubris, occurring outside the human boundary established by ancient tradition. The industrial economy is based on invasion and pillage of the Great Economy.

The weakness of the industrial economy is clearly revealed when it imposes its terms upon agriculture, for its terms cannot define those natural principles that are most vital to the life and longevity of farms. Even if the industrial economists could afford to do so, they could not describe the dependence of agriculture upon nature. If asked to consider the lilies of the field or told that the wheat is resurrected out of its graves, the agricultural industrialist would reply that "my engineer's mind inclines less toward the poetic and philosophical, and more toward the practical and possible," unable even to suspect that such a division of mind induces blindness to possibilities of the utmost practical concern.

That good topsoil both drains and retains water, that diversity increases capacity, are facts similarly alien to industrial logic. Industrialists see retention and drainage as different and opposite functions, and they would promote one at the expense of the other, just as, diversity being inimical to industrial procedure, they would commit themselves to the forlorn expedient of enlarging capacity by increasing area. They are thus encumbered by dependence on mechanical solutions that can work only by isolating and oversimplifying problems. Industrialists are condemned to proceed by devices. To facilitate water retention, they must resort to a specialized water-holding device such as a terrace or a dam; to facilitate drainage, they must use drain tile, or a ditch, or a "subsoiler." It is possible, I know, to argue that this analysis is too general and to produce exceptions, but I do not think it deniable that the discipline of soil conservation is now principally that of the engineer, not that of the farmer or soil husband—that it is now a matter of digging in the earth, not of enriching it.

I do not mean to say that the devices of engineering are always inappropriate; they have their place, not least in the restoration of land abused by the devices of engineering. My point is that, to facilitate both water retention and drainage in the same place, we must improve the soil, which is not a mechanical device but, among other things, a graveyard, a place of resurrection, and a community of living creatures. Devices may sometimes help, but only up to a point, for soil is improved by what humans do not do as well as by what they do. The proprieties of soil husbandry require acts that are much more complex than industrial acts, for these acts are conditioned by the ability *not* to act, by forbearance or self-restraint, sympathy or generosity. The industrial act is simply prescribed by thought, but the act of soil building is also *limited* by thought. We build soil by knowing what to do but also by knowing what not to do and by knowing when to stop. Both kinds of knowledge are necessary because invariably, at some point, the reach of human comprehension becomes too short, and at that point the work of the human economy must end in absolute deference to the working of the Great Economy. This, I take it, is the practical significance of the idea of the Sabbath.

To push our work beyond that point, invading the Great Economy, is to become guilty of hubris, of presuming to be greater than we are. We cannot do what the topsoil does, any more than we can do what God does or what a swallow does. We can fly, but only as humans — very crudely, noisily, and clumsily. We can dispose of corpses and garbage, but we cannot, by our devices, turn them into fertility and new life. And we are discovering, to our great uneasiness, that we cannot dispose at all of some of our so-called wastes that are toxic or radioactive. We can appropriate and in some fashion use godly powers, but we cannot use them safely, and we cannot control the results. That is to say that the human condition remains for us what it was for Homer and the authors of the Bible. Now that we have brought such enormous powers to our aid (we hope), it seems more necessary than ever to observe how inexorably the human condition still contains us. We only do what humans can do, and our machines, however they may appear to enlarge our possibilities, are invariably infected with our limitations. Sometimes, in enlarging our possibilities, they narrow our limits and leave us more powerful but less content, less safe, and less free. The mechanical means by which we propose to escape the human condition only extend it; thinking to transcend our definition as fallen creatures, we have only colonized more and more territory eastward of Eden.

II

Like the rich man of the parable, the industrialist thinks to escape the persistent obligations of the human condition by means of "much goods laid up for many years"— by means, in other words, of quantities: resources, supplies, stockpiles, funds, reserves. But this is a grossly oversimplifying dream and, thus, a dangerous one. All the great natural goods that empower agriculture, some of which I have discussed, have to do with quantities, but they have to do also with qualities, and they involve principles that are not static but active; they have to do with formal processes. The topsoil exists as such because it is ceaselessly transforming death into life, ceaselessly supplying food and water to all that lives in it and from it; otherwise, "All flesh shall perish together, and man shall turn again unto dust." If we are to live well on and from our land, we must live by faith in the ceaselessness of these processes and by faith in our own willingness and ability to collaborate with them. Christ's prayer for "daily bread" is an affirmation of such faith, just as it is a repudiation of faith in "much goods laid up." Our life and livelihood are the gift of the topsoil and of our willingness and ability to care for it, to grow good wheat, to make good bread; they do not derive from stockpiles of raw materials or accumulations of purchasing power.

The industrial economy can define potentiality, even the potentiality of the living topsoil, only as a *fund*, and thus it must accept impoverishment as the inescapable condition of abundance. The invariable mode of its relation both to nature and to human culture is that of mining: withdrawal from a limited fund until that fund is exhausted. It removes natural fertility and human workmanship from the land, just as it removes nourishment and human workmanship from bread. Thus the land is reduced to abstract marketable quantities of length and width, and bread to merchandise that is high in money value but low in food value. "Our bread," Guy Davenport once said to me, "is more obscene than our movies."

But the industrial use of *any* "resource" implies its exhaustion. It is for this reason that the industrial economy has been accompanied by an ever-increasing hurry of research and exploration, the motive of which is not "free enterprise" or "the spirit of free inquiry," as industrial scientists and apologists would have us believe, but the desperation that naturally and logically accompanies gluttony.

One of the favorite words of the industrial economy is "control": We want "to keep things under control"; we wish (or so we say) to "control" inflation and erosion; we have a discipline known as "crowd control"; we believe in "controlled growth" and "controlled development," in "traffic control" and "self-control." But, because we are always setting out to control something that we refuse to limit, we have made control a permanent and a helpless enterprise. If we will not limit causes, there can be no controlling of effects. What is to be the fate of self-control in an economy that encourages and rewards unlimited selfishness?

More than anything else, we would like to "control the forces of nature," refusing at the same time to impose any limit on human nature. We assume that such control and such freedom are our "rights," which seems to ensure that our means of control (of nature and of all else that we see as alien) will be violent. It is startling to recognize the extent to which the industrial economy depends upon controlled explosions—in mines, in weapons, in the cylinders of engines, in the economic pattern known as "boom and bust." This dependence is the result of a progress that can be argued for, but those who argue for it must recognize that, in all these means, good ends are served by a destructive principle, an association that is difficult to control if it is not limited; moreover, they must recognize that our failure to limit this association has raised the specter of uncontrollable explosion. Nuclear holocaust, if it comes, will be the final detonation of an explosive economy.

An explosive economy, then, is not only an economy that is dependent upon explosions but also one that sets no limits on itself. Any little economy that sees itself as unlimited is obviously self-blinded. It does not see its real relation of dependence and obligation to the Great Economy; in fact, it does not see that there *is* a Great Economy. Instead, it calls the Great Economy "raw material" or "natural resources" or "nature" and proceeds with the business of putting it "under control."

But "control" is a word more than ordinarily revealing here, for its root meaning is to roll against, in the sense of a little wheel turning in opposition. The principle of control, then, involves necessarily the principle of division: one thing may turn against another thing only by being divided from it. This mechanical division and turning in opposition William Blake understood as evil, and he spoke of "Satanic wheels" and "Satanic mills": "wheel without wheel, with cogs tyrannic/Moving by compulsion each other." By "wheel without wheel," Blake meant wheel outside of wheel, one wheel communicating motion to the other in the manner of two cogwheels, the point being that one wheel can turn another wheel outside itself only in a direction opposite to its own. This, I suppose, is acceptable enough as a mechanism. It becomes "Satanic" when it becomes a ruling metaphor and is used to describe and to organize fundamental relationships. Against the Satanic "wheel without wheel," Blake set the wheels of Eden, which "Wheel within wheel in freedom revolve, in harmony and peace." This is the "wheel in the middle of a wheel" of Ezekiel's vision, and it is an image of harmony. That the relation of these wheels is not mechanical we know from Ezekiel 1:21: "the spirit of the living creature was in the wheels." The wheels of opposition oppose the spirit of the living creature.

What had happened, as Blake saw accurately and feared justifiably, was a fundamental shift in the relation of humankind to the rest of creation. Sometime between, say, Pope's verses on the Chain of Being in *An Essay on Man* and Blake's "London," the dominant minds had begun to see the human race, not as a part or a member of Creation, but as outside it and opposed to it. The industrial revolution was only a part of this change, but it is true that, when the wheels of the industrial revolution began to revolve, they turned against nature, which became the name for all of Creation thought to be below humanity, as well as, incidentally, against all once thought to be above humanity. Perhaps this would have been safe enough if nature—that is, if all the rest of Creation—had been, as proposed, passively subject to human purpose.

Of course, it never has been. As Blake foresaw, and as we now know, what we turn against must turn against us. Blake's image of the cogwheels turning in relentless opposition is terrifyingly apt, for in our vaunted war against nature, nature fights back. The earth may answer our pinches and pokes "only with spring," as e. e. cummings said, but if we pinch and poke too much, she can answer also with flood or drouth, with catastrophic soil erosion, with plague and famine. Many of the occurrences that we call "acts of God" or "accidents of nature" are simply forthright natural responses to human provocations. Not always; I do not mean to imply here that, by living in harmony with nature, we can be free of floods and storms and drouth and earthquakes and volcanic eruptions; I am only pointing out, as many others have done, that, by living in opposition to nature, we can *cause* natural calamities of which we would otherwise be free.

* * *

The problem seems to be that a human economy cannot prescribe the terms of its own success. In a time when we wish to believe that humans are the sole authors of the truth, that truth is relative, and that value judgments are all subjective, it is hard to say that a human economy can be wrong, and yet we have good, sound, practical reasons for saying so. It is indeed possible for a human economy to be wrong—not relatively wrong, in the sense of being "out of adjustment," or unfair according to some human definition of fairness, or weak according to the definition of its own purposes—but wrong absolutely and according to practical measures. Of course, if we see the human economy as the *only* economy, we will see its errors as political failures, and we will continue to talk about "recovery." It is only when we think of the little human economy in relation to the Great Economy that we begin to understand our errors for what they are and to see the qualitative meanings of our quantitative measures. If we see the industrial economy in terms of the Great Economy, then we begin to see industrial wastes and losses not as "trade-offs" or "necessary risks" but as costs that, like all costs, are chargeable to somebody, sometime.

That we can prescribe the terms of our own success, that we can live outside or in ignorance of the Great Economy are the greatest errors. They condemn us to a life without a standard, wavering in inescapable bewilderment from paltry self-satisfaction to paltry self-dissatisfaction. But since we have no place to live but in the Great Economy, whether or not we know that and act accordingly is the critical question, not about economy merely, but about human life itself.

It is possible to make a little economy, such as our present one, that is so shortsighted and in which accounting is of so short a term as to give the impression that vices are necessary and practically justifiable. When we make our economy a little wheel turning in opposition to what we call "nature," then we set up competitiveness as the ruling principle in our explanation of reality and in our understanding of economy; we make of it, willy-nilly, a virtue. But competitiveness, as a ruling principle and a virtue, imposes a logic that is extremely difficult, perhaps impossible, to control. That logic explains why our cars and our clothes are shoddily made, why our "wastes" are toxic, and why our "defensive" weapons are suicidal; it explains why it is so difficult for us to draw a line between "free enterprise" and crime. If our economic ideal is maximum profit with minimum responsibility, why should we be surprised to find our corporations so frequently in court and robbery on the increase? Why should we be surprised to find that medicine has become an exploitive industry, profitable in direct proportion to its hurry and its mechanical indifference? People who pay for shoddy products or careless services and people who are robbed outright are equally victims of theft, the only difference being that the robbers outright are not guilty of fraud.

If, on the other hand, we see ourselves as living within the Great Economy, under the necessity of making our little human economy within it, according to its terms, the smaller wheel turning in sympathy with the greater, receiving its being and its motion from it, then we see that the traditional virtues are necessary and are practically justifiable. Then, because in the Great Economy *all* transactions count and the account is never "closed," the ideal changes. We see that we cannot *afford* maximum profit or power with minimum responsibility because, in the Great Economy, the loser's losses finally afflict the winner. Now the ideal must be "the maximum of well-being with the minimum of consumption," which both defines and requires neighborly love. Competitiveness cannot be the ruling principle, for the Great Economy is not a "side" that we can join nor are there such "sides" within it. Thus, it is not the "sum of its parts" but a *membership* of parts inextricably joined to each other, indebted to each other, receiving significance and worth from each other and from the whole. One is obliged to "consider the lilies of the field," not because they are lilies or because they are exemplary, but because they are fellow members and because, as fellow members, we and the lilies are in certain critical ways alike.

To say that within the Great Economy the virtues are necessary and practically justifiable is at once to remove them from that specialized, sanctimonious, condescending practice of virtuousness that is humorless, pointless, and intolerable to its beneficiaries. For a human, the good choice in the Great Economy is to see its membership as a neighborhood and oneself as a neighbor within it. I am sure that virtues count in a neighborhood—to "love thy neighbor as thyself" requires the help of all seven of them—but I am equally sure that in a neighborhood the virtues cannot be practiced as such. Temperance has no appearance or action of its own, nor does justice, prudence, fortitude, faith, hope, or charity. They can only be employed on occasions. "He who would do good to another," William Blake said, "must do it in Minute Particulars." To help each other, that is, we must go beyond the coldhearted charity of the "general good" and get down to work where we are:

> Labour well the Minute Particulars, attend to the Little-ones,
> And those who are in misery cannot remain so long
> If we do but our duty: labour well the teeming Earth.

It is the Great Economy, not any little economy, that invests minute particulars with high and final importance. In the Great Economy, each part stands for the whole and is joined to it; the whole is present in the part and is its health. The industrial economy, by contrast, is always striving and failing to make fragments (pieces that *it* has broken) *add up* to an ever-fugitive wholeness.

Work that is authentically placed and understood within the Great Economy moves virtue toward virtuosity—that is, toward skill or technical competence. There is no use in helping our neighbors with their work if we do not know how to work. When the virtues are rightly practiced within the Great Economy, we do not call them virtues; we call them good farming, good forestry, good carpentry, good husbandry, good weaving and sewing, good homemaking, good parenthood, good neighborhood, and so on. The general principles are submerged in the particularities of their engagement with the world. Lao Tzu saw the appearance of the virtues as such, in the abstract, as indicative of their loss:

> When people lost sight of the way to live
> Came codes of love and honesty . . .
> When differences weakened family ties
> Came benevolent fathers and dutiful sons;
> And when lands were disrupted and misgoverned
> Came ministers commended as loyal.

And these lines might be read as an elaboration of the warning against the *appearances* of goodness at the beginning of the sixth chapter of Matthew.

The work of the small economy, when it is understandingly placed within the Great Economy, minutely particularizes the virtues and carries principle into practice; to the extent that it does so, it escapes specialization. The industrial economy requires the extreme specialization of work — the separation of work from its results — because it subsists upon divisions of interest and must deny the fundamental kinships of producer and consumer; seller and buyer; owner and worker; worker, work, and product; parent material and product; nature and artifice; thoughts, words, and deeds. Divided from those kinships, specialized artists and scientists identify themselves as "observers" or "objective observers"— that is, as outsiders without responsibility or involvement. But the industrialized arts and sciences are false, their division is a lie, for there is no specialization of results.

There is no "outside" to the Great Economy, no escape into either specialization or generality, no "time off." Even insignificance is no escape, for in the membership of the Great Economy everything signifies; whatever we do counts. If we do not serve what coheres and endures, we serve what disintegrates and destroys. We can *presume* that we are outside the membership that includes us, but that presumption only damages the membership — and ourselves, of course, along with it.

In the industrial economy, the arts and the sciences are specialized "professions," each having its own language, speaking to none of the others. But the Great Economy proposes arts and sciences of membership: ways of doing and ways of knowing that cannot be divided from each other or within themselves and that speak the common language of the communities where they are practiced.

This modern mind sees only half of the horse — that half which may become a dynamo, or an automobile, or any other horsepowered machine. If this mind had much respect for the full-dimensioned, grass-eating horse, it would never have invented the engine which represents only half of him. The religious mind, on the other hand, has this respect; it wants the whole horse, and it will be satisfied with nothing less.

I should say a religious mind that requires more than a half-religion.

Allen Tate, "Remarks on the Southern Religion," in *I'll Take My Stand*

The Whole Horse

One of the primary results — and one of the primary needs — of industrialism is the separation of people and places and products from their histories. To the extent that we participate in the industrial economy, we do not know the histories of our families or of our habitats or of our meals. This is an economy, and in fact a culture, of the one-night stand. "I had a good time," says the industrial lover, "but don't ask me my last name." Just so, the industrial eater says to the svelte industrial hog, "We'll be together at breakfast. I don't want to see you before then, and I won't care to remember you afterwards."

In this condition, we have many commodities, but little satisfaction, little sense of the sufficiency of anything. The scarcity of satisfaction makes of our many commodities, in fact, an infinite series of commodities, the new commodities invariably promising greater satisfaction than the older ones. And so we can say that the industrial economy's most-marketed commodity is satisfaction, and that this commodity, which is repeatedly promised, bought, and paid for, is never delivered. On the other hand, people who have much satisfaction do not need many commodities.

The persistent want of satisfaction is directly and complexly related to the dissociation of ourselves and all our goods from our and their histories. If things do not last, are not made to last, they can have no histories, and we who use these things can have no memories. We buy new stuff on the promise of satisfaction because we have forgot the promised satisfaction for which we bought our old stuff. One of the procedures of the industrial economy is to reduce the longevity of materials. For example, wood, which well-made into buildings and furniture and well cared for can last hundreds of years, is now routinely manufactured into products that last twenty-five years. We do not cherish the memory of shoddy and transitory objects, and so we do not remember them. That is to say that we do not invest in them the lasting respect and admiration that make for satisfaction.

The problem of our dissatisfaction with all the things that we use is not correctable within the terms of the economy that produces those things. At present, it is virtually impossible for us to know the economic history or the ecological cost of the products we buy; the origins of the products are typically too distant and too scattered and the processes of trade, manufacture, transportation, and marketing too complicated. There are, moreover, too many good reasons for the industrial suppliers of these products not to want their histories to be known.

When there is no reliable accounting and therefore no competent knowledge of the economic and ecological effects of our lives, we cannot live lives that are economically and ecologically responsible. This is the problem that has frustrated, and to a considerable extent undermined, the American conservation effort from the beginning. It is ultimately futile to plead and protest and lobby in favor of public ecological responsibility while, in virtually every act of our private lives, we endorse and support an economic system that is by intention, and perhaps by necessity, ecologically irresponsible.

If the industrial economy is not correctable within or by its own terms, then obviously what is required for correction is a countervailing economic idea. And the most significant weakness of the conservation movement is its failure to produce or espouse an economic idea capable of correcting the economic idea of the industrialists. Somewhere near the heart of the conservation effort as we have known it is the romantic assumption that, if we have become alienated from nature, we can become unalienated by making nature the subject of contemplation or art, ignoring the fact that we live necessarily in and from nature—ignoring, in other words, all the economic issues that are involved. Walt Whitman could say, "I think I could turn and live with animals" as if he did not know that, in fact, we *do* live with animals, and that the terms of our relation to them are inescabably established by our economic use of their and our world. So long as we live, we are going to be living with skylarks, nightingales, daffodils, waterfowl, streams, forests, mountains, and all the other creatures that romantic poets and artists have yearned toward. And by the way we live we will determine whether or not those creatures will live.

That this nature-romanticism of the nineteenth century ignores economic facts and relationships has not prevented it from setting the agenda for modern conservation groups. This agenda has rarely included the economics of land use, without which the conservation effort becomes almost inevitably long on sentiment and short on practicality. The giveaway is that when conservationists try to be practical they are likely to defend the "sustainable use of natural resources" with the argument that this will make the industrial economy sustainable. A further giveaway is that the longer the industrial economy lasts in its present form, the further it will demonstrate its ultimate impossibility: every human in the world cannot, now or ever, own the whole catalogue of shoddy, high-energy industrial products which cannot be sustainably made or used. Moreover, the longer the industrial economy lasts, the more it will eat away the possibility of a better economy.

The conservation effort has at least brought under suspicion the general relativism of our age. Anybody who has studied with care the issues of conservation knows that our acts are being measured by a real and unyielding standard that was invented by no human. Our acts that are not in harmony with nature are inevitably and sometimes irremediably destructive. The standard exists. But having no opposing economic idea, conservationists have had great difficulty in applying the standard.

* * *

What, then, is the countervailing idea by which we might correct the industrial idea? We will not have to look hard to find it, for there is only one, and that is agrarianism. Our major difficulty (and danger) will be in attempting to deal with agrarianism as "an idea"—agrarianism is primarily a practice, a set of attitudes, a loyalty, and a passion; it is an idea only secondarily and at a remove. To use merely the handiest example: I was raised by agrarians, my bias and point of view from my earliest childhood were agrarian, and yet I never heard agrarianism defined, or even so much as named, until I was a sophomore in college. I am well aware of the danger in defining things, but if I am going to talk about agrarianism, I am going to have to define it. The definition that follows is derived both from agrarian writers, ancient and modern, and from the unliterary and sometimes illiterate agrarians who have been my teachers.

The fundamental difference between industrialism and agrarianism is this:

whereas industrialism is a way of thought based on monetary capital and technology, agrarianism is a way of thought based on land.

Agrarianism, furthermore, is a culture at the same time that it is an economy. Industrialism is an economy before it is a culture. Industrial culture is an accidental by-product of the ubiquitous effort to sell unnecessary products for more than they are worth.

An agrarian economy rises up from the fields, woods, and streams—from the complex of soils, slopes, weathers, connections, influences, and exchanges that we mean when we speak, for example, of the local community or the local watershed. The agrarian mind is therefore not regional or national, let alone global, but local. It must know on intimate terms the local plants and animals and local soils; it must know local possibilities and impossibilities, opportunities and hazards. It depends and insists on knowing very particular local histories and biographies.

Because a mind so placed meets again and again the necessity for work to be good, the agrarian mind is less interested in abstract quantities than in particular qualities. It feels threatened and sickened when it hears people and creatures and places spoken of as labor, management, capital, and raw material. It is not at all impressed by the industrial legendry of gross national products, or of the numbers sold and dollars earned by gigantic corporations. It is interested—and forever fascinated—by questions leading toward the accomplishment of good work: What is the best location for a particular building or fence? What is the best way to plow *this* field? What is the best course for a skid road in *this* woodland? Should *this* tree be cut or spared? What are the best breeds and types of livestock for *this* farm?—questions which cannot be answered in the abstract, and which yearn not toward quantity but toward elegance. Agrarianism can never become abstract because it has to be practiced in order to exist.

And though this mind is local, almost absolutely placed, little attracted to mobility either upward or lateral, it is not provincial; it is too taken up and fascinated by its work to feel inferior to any other mind in any other place.

An agrarian economy is always a subsistence economy before it is a market economy. The center of an agrarian farm is the household. The function of the household economy is to assure that the farm family lives as far as possible from the farm. It is the subsistence part of the agrarian economy that assures its stability and its survival. A subsistence economy necessarily is highly diversified, and it characteristically has involved hunting and gathering as well as farming and gardening. These activities bind people to their local landscape by close, complex interests and economic ties. The industrial economy alienates people from the native landscape precisely by breaking these direct practical ties and introducing distant dependences.

Agrarian people of the present, knowing that the land must be well cared for if anything is to last, understand the need for a settled connection, not just between farmers and their farms, but between urban people and their surrounding and tributary landscapes. Because the knowledge and know-how of good caretaking must be handed down to children, agrarians recognize the necessity of preserving the coherence of families and communities.

The stability, coherence, and longevity of human occupation require that the land should be divided among many owners and users. The central figure of agrarian thought has invariably been the small owner or small holder who maintains a significant measure of economic self-determination on a small acreage. The scale and independence of such holdings imply two things that agrarians see as desirable: intimate care in the use of the land, and political democracy resting upon the indispensable foundation of economic democracy.

A major characteristic of the agrarian mind is a longing for independence —that is, for an appropriate degree of personal and local self-sufficiency. Agrarians wish to earn and deserve what they have. They do not wish to live by piracy, beggary, charity, or luck.

In the written record of agrarianism, there is a continually recurring affirmation of nature as the final judge, law-giver, and pattern-maker of and for the human use of the earth. We can trace the lineage of this thought in the West through the writings of Virgil, Spenser, Shakespeare, Pope, Thomas Jefferson, and on into the work of the twentieth-century agriculturists and scientists, J. Russell Smith, Liberty Hyde Bailey, Albert Howard, Wes Jackson, John Todd, and others. The idea is variously stated: we should not work until we have looked and seen where we are; we should honor Nature not only as our mother or grandmother, but as our teacher and judge; we should "let the forest judge"; we should "consult the Genius of the Place": we should make the farming fit the farm; we should carry over into the cultivated field the diversity and coherence of the native forest or prairie. And this way of thinking is surely allied to that of the medieval scholars and architects who saw the building of a cathedral as a symbol or analogue of the creation of the world. The agrarian mind is, at bottom, a religious mind. It subscribes to Allen Tate's doctrine of "the whole horse." It prefers the Creation itself to the powers and quantities to which it can be reduced. And this is a mind completely different from that which sees creatures as machines, minds as computers, soil fertility as chemistry, or agrarianism as an idea. John Haines has written that "the eternal task of the artist and the poet, the historian and the scholar . . . is to find the means to reconcile what are two separate and yet inseparable histories, Nature and Culture. To the extent that we can do this, the 'world' makes sense to us and can be lived in." I would add only that this applies also to the farmer, the forester, the scientist, and others.

The agrarian mind begins with the love of fields and ramifies in good farming, good cooking, good eating, and gratitude to God. Exactly analogous to the agrarian mind is the sylvan mind that begins with the love of forests and ramifies in good forestry, good woodworking, good carpentry, etc., and gratitude to God. These two kinds of mind readily intersect and communicate; neither ever intersects or communicates with the industrial-economic mind. The industrial-economic mind begins with ingratitude, and ramifies in the destruction of farms and forests. The "lowly" and "menial" arts of farm and forest are mostly taken for granted or ignored by the culture of the "fine arts" and by "spiritual" religions; they are taken for granted or ignored or held in contempt by the powers of the industrial economy. But in fact they are inescapably the foundation of human life and culture, and their adepts are capable of as deep satisfactions and as high attainments as anybody else.

Having, so to speak, laid industrialism and agrarianism side by side, implying a preference for the latter, I will be confronted by two questions that I had better go ahead and answer.

The first is whether or not agrarianism is simply a "phase" that we humans had to go through and then leave behind in order to get onto the track of technological progress toward ever greater happiness. The answer is that although industrialism has certainly conquered agrarianism, and has very nearly destroyed it altogether, it is also true that in every one of its uses of the natural world industrialism is in the process of catastrophic failure. Industry is now desperately shifting—by means of genetic engineering, global colonialism, and other contrivances—to prolong its control of our farms and forests, but the failure nonetheless continues. It is not possible to argue sanely in favor of soil erosion, water pollution, genetic impoverishment, and the destruction of rural communities and local economies. Industrialism, unchecked by the affections and concerns of agrarianism, becomes monstrous. And this is because of a weakness identified by the Twelve Southerners of *I'll Take My Stand* in their "Statement of Principles": under the rule of industrialism "the remedies proposed . . . are always homeopathic." Industrialism always proposes to correct its errors and excesses by more industrialization.

The second question is whether or not by espousing the revival of agrarianism we will commit the famous sin of "turning back the clock." The answer to that, for present-day North Americans, is fairly simple. The overriding impulse of agrarianism is toward the local adaptation of economies and cultures. Agrarian people wish to fit the farming to the farm and the forestry to the forest. At times and in places we latter-day Americans may have come close to accomplishing this goal, and we have a few surviving examples, but it is generally true that we are much farther from local adaptation now than we were fifty years ago. We never yet have developed stable, sustainable, locally adapted land-based economies. The good rural enterprises and communities that we will find in our past have been almost constantly under threat from the colonialism, first foreign and then domestic and now "global," which has so far dominated our history, and which has been institutionalized for a long time in the industrial economy. The possibility of an authentically settled country still lies ahead of us.

* * *

If we wish to look ahead, we will see not only in the United States but in the world two economic programs that conform pretty exactly to the aims of industrialism and agrarianism as I have described them.

The first is the effort to globalize the industrial economy, not merely by the expansionist programs of supra-national corporations within themselves, but also by means of government-sponsored international trade agreements, the most prominent of which is the World Trade Organization agreement, which institutionalizes the industrial ambition to use, sell, or destroy every acre and every creature of the world.

The World Trade Organization gives the lie to the industrialist conservatives' professed abhorrence of big government. The cause of big government, after all, is big business. The power to do large-scale damage, which is gladly assumed by every large-scale industrial enterprise, calls naturally and logically for government regulation, which of course the corporations object to. But we have a good deal of evidence also that the leaders of big business actively desire and promote big government. They and their political allies, while ostensibly working to "downsize" government, continue to promote government helps and "incentives" to large corporations, and, however absurdly, to adhere to their notion that a small government, taxing only the working people, can maintain a big highway system, a big military establishment, and a big space program, and award big government contracts.

But the most damaging evidence is the World Trade Organization itself, which is in effect a global government, with power to enforce the decisions of the collective against national laws that conflict with it. The coming of the World Trade Organization was foretold seventy years ago in the "Statement of Principles" of *I'll Take My Stand*, which said that "the true Sovietists or Communists . . . are the industrialists themselves. They would have the government set up an economic super-organization, which in turn would become the government." The agrarians of *I'll Take My Stand* did not foresee this because they were fortune-tellers, but because they had perceived accurately the character and motive of the industrial economy.

The second program, counter to the first, is composed of many small efforts to preserve or improve or establish local economies. These efforts on the part of nonindustrial or agrarian conservatives, local patriots, are taking place in countries both affluent and poor all over the world.

Whereas the corporate sponsors of the World Trade Organization, in order to promote their ambitions, have required only the hazy glamour of such phrases as "the global economy," "the global context," and "globalization"—and thus apparently have vacuum-packed the mind of every politician and political underling in the world—the local economists use a much more diverse and particularizing vocabulary that you can actually think with: "community," "ecosystem," "watershed," "place," "homeland," "family," "household."

And whereas the global economists advocate a world-government-by-economic-bureaucracy, which would destroy local adaptation everywhere by ignoring the uniqueness of every place, the local economists found their work upon respect for such uniqueness. Places differ from one another, the local economists say, therefore we must behave with unique consideration in each one; the ability to tender an appropriate practical regard and respect to each place in its difference is a kind of freedom; the inability to do so is a kind of tyranny. The global economists are the great centralizers of our time. The local economists, who have so far attracted the support of no prominent politician, are the true decentralizers and downsizers, for they seek an appropriate degree of self-determination and independence for localities. They seem to be moving toward a radical and necessary revision of our idea of a city. They are learning to see the city, not just as a built and paved municipality set apart by "city limits" to live by trade and transportation from the world at large, but rather as a part of a community which includes also the city's rural neighbors, its surrounding landscape and its watershed, on which it might depend for at least some of its necessities, and for the health of which it might exercise a competent concern and responsibility.

At this point, I want to say point-blank what I hope is already clear: though agrarianism proposes that everybody has agrarian responsibilities, it does not propose that everybody should be a farmer or that we do not need cities. Nor does it propose that every product should be a necessity. Furthermore, any thinkable human economy would have to grant to manufacturing an appropriate and honorable place. Agrarians would insist only that any manufacturing enterprise should be formed and scaled to fit the local landscape, the local ecosystem, and the local community, and that it should be locally owned and employ local people. They would insist, in other words, that the shop or factory owner should not be an outsider, but rather a sharer in the fate of the place and its community. The deciders should live with the results of their decisions.

Between these two programs—the industrial and the agrarian, the global and the local—the most critical difference is that of knowledge. The global economy institutionalizes a global ignorance, in which producers and consumers cannot know or care about one another, and in which the histories of all products will be lost. In such a circumstance, the degradation of products and places, producers and consumers, is inevitable.

But in a sound local economy, in which producers and consumers are neighbors, nature will become the standard of work and production. Consumers who understand their economy will not tolerate the destruction of the local soil or ecosystem or watershed as a cost of production. Only a healthy local economy can keep nature and work together in the consciousness of the community. Only such a community can restore history to economics.

* * *

I will not be altogether surprised to be told that I have set forth here a line of thought that is attractive but hopeless. A number of critics have advised me of this, out of their charity, as if I might have written of my hopes for forty years without giving a thought to hopelessness. Hope, of course, is always accompanied by the fear of hopelessness, which is a legitimate fear.

And so I would like to conclude by confronting directly the issue of hope. My hope is most seriously challenged by the fact of decline, of loss. The things that I have tried to defend are less numerous and worse off now than when I started, but in this I am only like all other conservationists. All of us have been fighting a battle that on average we are losing, and I doubt that there is any use in reviewing the statistical proofs. The point—the only interesting point—is that we have not quit. Ours is not a fight that you can stay in very long if you look on victory as a sign of triumph or on loss as a sign of defeat. We have not quit because we are not hopeless.

My own aim is not hopelessness. I am not looking for reasons to give up. I am looking for reasons to keep on. In outlining here the concerns of agrarianism, I have intended to show how the effort of conservation could be enlarged and strengthened.

What agrarian principles implicitly propose—and what I explicitly propose in advocating those principles at this time—is a revolt of local small producers and local consumers against the global industrialism of the corporations. Do I think that there is a hope that such a revolt can survive and succeed, and that it can have a significant influence upon our lives and our world?

Yes, I do. And to be as plain as possible, let me just say what I know. I know from friends and neighbors and from my own family that it is now possible for farmers to sell at a premium to local customers such products as organic vegetables, organic beef and lamb, and pasture-raised chickens. This market is being made by the exceptional goodness and freshness of the food, by the wish of urban consumers to support their farming neighbors, and by the excesses and abuses of the corporate food industry.

This, I think, gives the pattern of an economic revolt that is not only possible but is happening. It is happening for two reasons: First, as the scale of industrial agriculture increases, so does the scale of its abuses, and it is hard to hide large-scale abuses from consumers. It is virtually impossible now for intelligent consumers to be ignorant of the heartlessness and nastiness of animal confinement operations and their excessive use of antibiotics, of the use of hormones in meat and milk production, of the stenches and pollutants of pig and poultry factories, of the use of toxic chemicals and the waste of soil and soil health in industrial row-cropping, of the mysterious or disturbing or threatening practices associated with industrial food storage, preservation, and processing. Second, as the food industries focus more and more on gigantic global opportunities, they cannot help but overlook small local opportunities, as is made plain by the increase of "community-supported agriculture," farmers markets, health food stores, and so on. In fact, there are some markets that the great corporations by definition cannot supply. The market for so-called organic food, for example, is really a market for good, fresh, trustworthy food, food from producers known and trusted by consumers, and such food cannot be produced by a global corporation.

But the food economy is only one example. It is also possible to think of good local forest economies. And in the face of much neglect, it is possible to think of local small business economies—some of them related to the local economies of farm and forest—supported by locally owned, community-oriented banks.

What do these struggling, sometimes failing, sometimes hardly realized efforts of local economy have to do with conservation as we know it? The answer, probably, is *everything*. The conservation movement, as I said earlier, has a conservation program; it has a preservation program; it has a rather sporadic health-protection program; but it has no economic program, and because it has no economic program it has the status of something exterior to daily life, surviving by emergency, like an ambulance service. In saying this, I do not mean to belittle the importance of protest, litigation, lobbying, legislation, large-scale organization—all of which I believe in and support. I am saying simply that we must do more. We must confront, on the ground, and each of us at home, the economic assumptions in which the problems of conservation originate.

We have got to remember that the great destructiveness of the industrial age comes from a division, a sort of divorce, in our economy, and therefore in our consciousness, between production and consumption. Of this radical division of functions we can say, without much fear of oversimplifying, that the aim of producers is to sell as much as possible and that the aim of consumers is to buy as much as possible. We need only to add that the aim of both producer and consumer is to be so far as possible carefree. Because of various pressures, governments have learned to coerce from producers some grudging concern for the health and solvency of consumers. No way has been found to coerce from consumers any consideration for the methods and sources of production.

What alerts consumers to the outrages of producers is typically some kind of loss or threat of loss. We see that in dividing consumption from production we have lost the function of conserving. Conserving is no longer an integral part of the economy of the producer or the consumer. Neither the producer nor the consumer any longer says, "I must be careful of this so that it will last." The working assumption of both is that where there is some, there must be more. If they can't get what they need in one place, they will find it in another. That is why conservation is now a separate concern, a separate effort.

But experience seems increasingly to be driving us out of the categories of producer and consumer and into the categories of citizen, family member, and community member, in all of which we have an inescapable interest in making things last. And here is where I think the conservation movement (I mean that movement that has defined itself as the defender of wilderness and the natural world) can involve itself in the fundamental issues of economy and land use, and in the process gain strength for its original causes.

I would like my fellow conservationists to notice how many people and organizations are now working to save something of value—not just wilderness places, wild rivers, wildlife habitat, species diversity, water quality, and air quality, but also agricultural land, family farms and ranches, communities, children and childhood, local schools, local economies, local food markets, livestock breeds and domestic plant varieties, fine old buildings, scenic roads, and so on. I would like my fellow conservationists to understand also that there is hardly a small farm or ranch or locally owned restaurant or store or shop or business anywhere that is not struggling to save itself.

All of these people, who are fighting sometimes lonely battles to preserve things of value that they cannot bear to lose, are the conservation movement's natural allies. Most of them have the same enemies as the conservation movement. There is no necessary conflict among them. Thinking of them, in their great variety, in the essential likeness of their motives and concerns, one thinks of the possibility of a defined community of interest among them all, a shared stewardship of all the diversity of good things that are needed for the health and abundance of the world.

I don't suppose that this will be easy, given especially the history of conflict between conservationists and land users. I only suppose that it is necessary. Conservationists can't conserve everything that needs conserving without joining the effort to use well the agricultural lands, the forests, and the waters that we must use. To enlarge the areas protected from use without at the same time enlarging the areas of *good* use is a mistake. To have no large areas of protected old-growth forest would be folly, as most of us would agree. But it is also folly to have come this far in our history without a single working model of a thoroughly diversified and integrated, ecologically sound, local forest economy. That such an economy is possible is indicated by many imperfect or incomplete examples, but we need desperately to put the pieces together in one place—and then in every place.

The most tragic conflict in the history of conservation is that between the conservationists and the farmers and ranchers. It is tragic because it is unnecessary. There is no irresolvable conflict here, but the conflict that exists can be resolved only on the basis of a common understanding of good practice. Here again we need to foster and study working models: farms and ranches that are knowledgeably striving to bring economic practice into line with ecological reality, and local food economies in which consumers conscientiously support the best land stewardship.

We know better than to expect very soon a working model of a conserving global corporation. But we must begin to expect—and we must, as conservationists, begin working for, and in—working models of conserving local economies. These are possible now. Good and able people are working hard to develop them now. They need the full support of the conservation movement now. Conservationists need to go to these people, ask what they can do to help, and then help. A little later, having helped, they can in turn ask for help.

The Idea of a Local Economy

Let us begin by assuming what appears to be true: that the so-called environmental crisis is now pretty well established as a fact of our age. The problems of pollution, species extinction, loss of wilderness, loss of farmland, loss of topsoil may still be ignored or scoffed at, but they are not denied. Concern for these problems has acquired a certain standing, a measure of discussability, in the media and in some scientific, academic, and religious institutions.

This is good, of course; obviously, we can't hope to solve these problems without an increase of public awareness and concern. But in an age burdened with "publicity," we have to be aware also that as issues rise into popularity they rise also into the danger of oversimplification. To speak of this danger is especially necessary in confronting the destructiveness of our relationship to nature, which is the result, in the first place, of gross oversimplification.

The "environmental crisis" has happened because the human household or economy is in conflict at almost every point with the household of nature. We have built our household on the assumption that the natural household is simple and can be simply used. We have assumed increasingly over the last five hundred years that nature is merely a supply of "raw materials," and that we may safely possess those materials merely by taking them. This taking, as our technical means have increased, has involved always less reverence or respect, less gratitude, less local knowledge, and less skill. Our methodologies of land use have strayed from our old sympathetic attempts to imitate natural processes, and have come more and more to resemble the methodology of mining, even as mining itself has become more technologically powerful and more brutal.

And so we will be wrong if we attempt to correct what we perceive as "environmental" problems without correcting the economic oversimplification that caused them. This oversimplification is now either a matter of corporate behavior or of behavior under the influence of corporate behavior. This is sufficiently clear to many of us. What is not sufficiently clear, perhaps to any of us, is the extent of our complicity, as individuals and especially as individual consumers, in the behavior of the corporations.

What has happened is that most people in our country, and apparently most people in the "developed" world, have given proxies to the corporations to produce and provide *all* of their food, clothing, and shelter. Moreover, they are rapidly giving proxies to corporations or governments to provide entertainment, education, child care, care of the sick and the elderly, and many other kinds of "service" that once were carried on informally and inexpensively by individuals or households or communities. Our major economic practice, in short, is to delegate the practice to others.

The danger now is that those who are concerned will believe that the solution to the "environmental crisis" can be merely political—that the problems, being large, can be solved by large solutions generated by a few people to whom we will give our proxies to police the economic proxies that we have already given. The danger, in other words, is that people will think they have made a sufficient change if they have altered their "values," or had a "change of heart," or experienced a "spiritual awakening," and that such a change in passive consumers will cause appropriate changes in the public experts, politicians, and corporate executives to whom they have granted their political and economic proxies.

The trouble with this is that a proper concern for nature and our use of nature must be practiced not by our proxy-holders, but by ourselves. A change of heart or of values without a practice is only another pointless luxury of a passively consumptive way of life. The "environmental crisis," in fact, can be solved only if people, individually and in their communities, recover responsibility for their thoughtlessly given proxies. If people begin the effort to take back into their own power a significant portion of their economic responsibility, then their inevitable first discovery is that the "environmental crisis" is no such thing; it is not a crisis of our environs or surroundings; it is a crisis of our lives as individuals, as family members, as community members, and as citizens. We have an "environmental crisis" because we have consented to an economy in which by eating, drinking, working, resting, traveling, and enjoying ourselves we are destroying the natural, the god-given world.

* * *

We live, as we must sooner or later recognize, in an era of sentimental economics and, consequently, of sentimental politics. Sentimental communism holds in effect that everybody and everything should suffer for the good of "the many" who, though miserable in the present, will be happy in the future for exactly the same reasons that they are miserable in the present.

Sentimental capitalism is not as different from sentimental communism as the corporate and political powers claim. Sentimental capitalism holds in effect that everything small, local, private, personal, natural, good, and beautiful must be sacrificed in the interest of the "free market" and the great corporations, which will bring unprecedented security and happiness to "the many"—in, of course, the future.

These forms of political economy may be described as sentimental because they depend absolutely upon a political faith for which there is no justification, and because they issue a cold check on the virtue of political and/or economic rulers. They seek, that is, to preserve the gullibility of the people by appealing to a fund of political virtue that does not exist. Communism and "free-market" capitalism both are modern versions of oligarchy. In their propaganda, both justify violent means by good ends, which always are put beyond reach by the violence of the means. The trick is to define the end vaguely—"the greatest good of the greatest number" or "the benefit of the many"—and keep it at a distance.

The fraudulence of these oligarchic forms of economy is in their principle of displacing whatever good they recognize (as well as their debts) from the present to the future. Their success depends upon persuading people, first, that whatever they have now is no good, and second, that the promised good is certain to be achieved in the future. This obviously contradicts the principle—common, I believe, to all the religious traditions—that if ever we are going to do good to one another, then the time to do it is now; we are to receive no reward for promising to do it in the future. And both communism and capitalism have found such principles to be a great embarrassment. If you are presently occupied in destroying every good thing in sight in order to do good in the future, it is inconvenient to have people saying things like "Love thy neighbor as thyself" or "Sentient beings are numberless, I vow to save them." Communists and capitalists alike, "liberal" and "conservative" capitalists alike, have needed to replace religion with some form of determinism, so that they can say to their victims, "I am doing this because I can't do otherwise. It is not my fault. It is inevitable." The wonder is how often organized religion has gone along with this lie.

The idea of an economy based upon several kinds of ruin may seem a contradiction in terms, but in fact such an economy is possible, as we see. It is possible, however, on one implacable condition: the only future good that it assuredly leads to is that it will destroy itself. And how does it disguise this outcome from its subjects, its short-term beneficiaries, and its victims? It does so by false accounting. It substitutes for the real economy, by which we build and maintain (or do not maintain) our household, a symbolic economy of money, which in the long run, because of the self-interested manipulations of the "controlling interests," cannot symbolize or account for anything but itself. And so we have before us the spectacle of unprecedented "prosperity" and "economic growth" in a land of degraded farms, forests, ecosystems, and watersheds, polluted air, failing families, and perishing communities.

* * *

This moral and economic absurdity exists for the sake of the allegedly "free" market, the single principle of which is this: commodities will be produced wherever they can be produced at the lowest cost, and consumed wherever they will bring the highest price. To make too cheap and sell too high has always been the program of industrial capitalism. The idea of the global "free market" is merely capitalism's so-far-successful attempt to enlarge the geographic scope of its greed, and moreover to give to its greed the status of a "right" within its presumptive territory. The global "free market" is free to the corporations precisely because it dissolves the boundaries of the old national colonialisms, and replaces them with a new colonialism without restraints or boundaries. It is pretty much as if all the rabbits have now been forbidden to have holes, thereby "freeing" the hounds.

The "right" of a corporation to exercise its economic power without restraint is construed, by the partisans of the "free market," as a form of freedom, a political liberty implied presumably by the right of individual citizens to own and use property.

But the "free market" idea introduces into government a sanction of an inequality that is not implicit in any idea of democratic liberty: namely that the "free market" is freest to those who have the most money, and is not free at all to those with little or no money. Wal-Mart, for example, as a large corporation "freely" competing against local, privately owned businesses has virtually all the freedom, and its small competitors virtually none.

To make too cheap and sell too high, there are two requirements. One is that you must have a lot of consumers with surplus money and unlimited wants. For the time being, there are plenty of these consumers in the "developed" countries. The problem, for the time being easily solved, is simply to keep them relatively affluent and dependent on purchased supplies.

The other requirement is that the market for labor and raw materials should remain depressed relative to the market for retail commodities. This means that the supply of workers should exceed demand, and that the land-using economy should be allowed or encouraged to overproduce.

To keep the cost of labor low, it is necessary first to entice or force country people everywhere in the world to move into the cities—in the manner prescribed by the United States' Committee for Economic Development after World War II—and second, to continue to introduce labor-replacing technology. In this way it is possible to maintain a "pool" of people who are in the threatful position of being mere consumers, landless and also poor, and who therefore are eager to go to work for low wages—precisely the condition of migrant farm workers in the United States.

To cause the land-using economies to overproduce is even simpler. The farmers and other workers in the world's land-using economies, by and large, are not organized. They are therefore unable to control production in order to secure just prices. Individual producers must go individually to the market and take for their produce simply whatever they are paid. They have no power to bargain or make demands. Increasingly, they must sell, not to neighbors or to neighboring towns and cities, but to large and remote corporations. There is no competition among the buyers (supposing there is more than one), who *are* organized, and are "free" to exploit the advantage of low prices. Low prices encourage overproduction as producers attempt to make up their losses "on volume," and overproduction inevitably makes for low prices. The land-using economies thus spiral downward as the money economy of the exploiters spirals upward. If economic attrition in the land-using population becomes so severe as to threaten production, then governments can subsidize production without production controls, which necessarily will encourage overproduction, which will lower prices — and so the subsidy to rural producers becomes, in effect, a subsidy to the purchasing corporations. In the land-using economies production is further cheapened by destroying, with low prices and low standards of quality, the cultural imperatives for good work and land stewardship.

* * *

This sort of exploitation, long familiar in the foreign and domestic economies and the colonialism of modern nations, has now become "the global economy," which is the property of a few supranational corporations. The economic theory used to justify the global economy in its "free market" version is again perfectly groundless and sentimental. The idea is that what is good for the corporations will sooner or later — though not of course immediately — be good for everybody.

That sentimentality is based in turn upon a fantasy: the proposition that the great corporations, in "freely" competing with one another for raw materials, labor, and marketshare, will drive each other indefinitely, not only toward greater "efficiencies" of manufacture, but also toward higher bids for raw materials and labor and lower prices to consumers. As a result, all the world's people will be economically secure — in the future. It would be hard to object to such a proposition if only it were true.

But one knows, in the first place, that "efficiency" in manufacture always means reducing labor costs by replacing workers with cheaper workers or with machines.

In the second place, the "law of competition" does *not* imply that many competitors will compete indefinitely. The law of competition is a simple paradox: competition destroys competition. The law of competition implies that many competitors, competing without restraint, will ultimately and inevitably reduce the number of competitors to one. The law of competition, in short, is the law of war.

In the third place, the global economy is based upon cheap long-distance transportation, without which it is not possible to move goods from the point of cheapest origin to the point of highest sale. And cheap long-distance transportation is the basis of the idea that regions and nations should abandon any measure of economic self-sufficiency in order to specialize in production for export of the few commodities or the single commodity that can be most cheaply produced. Whatever may be said for the "efficiency" of such a system, its result (and I assume, its purpose) is to destroy local production capacities, local diversity, and local economic independence.

This idea of a global "free market" economy, despite its obvious moral flaws and its dangerous practical weaknesses, is now the ruling orthodoxy of the age. Its propaganda is subscribed to and distributed by most political leaders, editorial writers, and other "opinion makers." The powers that be, while continuing to budget huge sums for "national defense," have apparently abandoned any idea of national or local self-sufficiency, even in food. They also have given up the idea that a national or local government might justly place restraints upon economic activity in order to protect its land and its people.

The global economy is now institutionalized in the World Trade Organization, which was set up, without election anywhere, to rule international trade on behalf of the "free market"—which is to say on behalf of the supranational corporations—and to *over*rule, in secret sessions, any national or regional law that conflicts with the "free market." The corporate program of global free trade and the presence of the World Trade Organization have legitimized extreme forms of expert thought. We are told confidently that if Kentucky loses its milk-producing capacity to Wisconsin, that will be a "success story." Experts such as Stephen C. Blank of the University of California, Davis, have proposed that "developed" countries, such as the United States and the United Kingdom, where food can no longer be produced cheaply enough, should give up agriculture altogether.

The folly at the root of this foolish economy began with the idea that a corporation should be regarded, legally, as "a person." But the limitless destructiveness of this economy comes about precisely because a corporation is *not* a person. A corporation, essentially, is a pile of money to which a number of persons have sold their moral allegiance. As such, unlike a person, a corporation does not age. It does not arrive, as most persons finally do, at a realization of the shortness and smallness of human lives; it does not come to see the future as the lifetimes of the children and grandchildren of anybody in particular. It can experience no personal hope or remorse, no change of heart. It cannot humble itself. It goes about its business as if it were immortal, with the single purpose of becoming a bigger pile of money. The stockholders essentially are usurers, people who "let their money work for them," expecting high pay in return for causing others to work for low pay. The World Trade Organization enlarges the old idea of the corporation-as-person by giving the global corporate economy the status of a super government with the power to overrule nations.

I don't mean to say, of course, that all corporate executives and stockholders are bad people. I am only saying that all of them are very seriously implicated in a bad economy.

* * *

Unsurprisingly, among people who wish to preserve things other than money—for instance, every region's native capacity to produce essential goods—there is a growing perception that the global "free market" economy is inherently an enemy to the natural world, to human health and freedom, to industrial workers, and to farmers and others in the land-use economies; and furthermore, that it is inherently an enemy to good work and good economic practice.

I believe that this perception is correct and that it can be shown to be correct merely by listing the assumptions implicit in the idea that corporations should be "free" to buy low and sell high in the world at large. These assumptions, so far as I can make them out, are as follows:

1. That stable and preserving relationships among people, places, and things do not matter and are of no worth.
2. That cultures and religions have no legitimate practical or economic concerns.
3. That there is no conflict between the "free market" and political freedom, and no connection between political democracy and economic democracy.
4. That there can be no conflict between economic advantage and economic justice.
5. That there is no conflict between greed and ecological or bodily health.
6. That there is no conflict between self-interest and public service.
7. That the loss or destruction of the capacity anywhere to produce necessary goods does not matter and involves no cost.
8. That it is all right for a nation's or a region's subsistence to be foreign based, dependent on long-distance transport, and entirely controlled by corporations.
9. That, therefore, wars over commodities—our recent Gulf War, for example—are legitimate and permanent economic functions.

10. That this sort of sanctioned violence is justified also by the predominance of centralized systems of production supply, communications, and transportation, which are extremely vulnerable not only to acts of war between nations, but also to sabotage and terrorism.
11. That it is all right for poor people in poor countries to work at poor wages to produce goods for export to affluent people in rich countries.
12. That there is no danger and no cost in the proliferation of exotic pests, weeds, and diseases that accompany international trade and that increase with the volume of trade.
13. That an economy is a machine, of which people are merely the interchangeable parts. One has no choice but to do the work (if any) that the economy prescribes, and to accept the prescribed wage.
14. That, therefore, vocation is a dead issue. One does not do the work that one chooses to do because one is called to it by Heaven or by one's natural or god-given abilities, but does instead the work that is determined and imposed by the economy. Any work is all right as long as one gets paid for it.

These assumptions clearly prefigure a condition of total economy. A total economy is one in which everything—"life forms," for instance, or the "right to pollute"—is "private property" and has a price and is for sale. In a total economy significant and sometimes critical choices that once belonged to individuals or communities become the property of corporations. A total economy, operating internationally, necessarily shrinks the powers of state and national governments, not only because those governments have signed over significant powers to an international bureaucracy or because political leaders become the paid hacks of the corporations but also because political processes—and especially democratic processes—are too slow to react to unrestrained economic and technological development on a global scale. And when state and national governments begin to act in effect as agents of the global economy, selling their people for low wages and their people's products for low prices, then the rights and liberties of citizenship must necessarily shrink. A total economy is an unrestrained taking of profits from the disintegration of nations, communities, households, landscapes, and ecosystems. It licenses symbolic or artificial wealth to "grow" by means of the destruction of the real wealth of all the world.

Among the many costs of the total economy, the loss of the principle of vocation is probably the most symptomatic and, from a cultural standpoint, the most critical. It is by the replacement of vocation with economic determinism that the exterior workings of a total economy destroy the character and culture also from the inside.

In an essay on the origin of civilization in traditional cultures, Ananda K. Coomaraswamy wrote that "the principle of justice is the same throughout . . . [it is] that each member of the community should perform the task for which he is fitted by nature. . . ." The two ideas, justice and vocation, are inseparable. That is why Coomaraswamy spoke of industrialism as "the mammon of injustice," incompatible with civilization. It is by way of the principle and practice of vocation that sanctity and reverence enter into the human economy. It was thus possible for traditional cultures to conceive that "to work is to pray."

* * *

Aware of industrialism's potential for destruction, as well as the considerable political danger of great concentrations of wealth and power in industrial corporations, American leaders developed, and for a while used, the means of limiting and restraining such concentrations, and of somewhat equitably distributing wealth and property. The means were: laws against trusts and monopolies, the principle of collective bargaining, the concept of one-hundred-percent parity between the land-using and the manufacturing economies, and the progressive income tax. And to protect domestic producers and production capacities it is possible for governments to impose tariffs on cheap imported goods. These means are justified by the government's obligation to protect the lives, livelihoods, and freedoms of its citizens. There is, then, no necessity or inevitability requiring our government to sacrifice the livelihoods of our small farmers, small business people, and workers, along with our domestic economic independence to the global "free market." But now all of these means are either weakened or in disuse. The global economy is intended as a means of subverting them.

In default of government protections against the total economy of the supranational corporations, people are where they have been many times before: in danger of losing their economic security and their freedom, both at once. But at the same time the means of defending themselves belongs to them in the form of a venerable principle: powers not exercised by government return to the people. If the government does not propose to protect the lives, livelihoods, and freedoms of its people, then the people must think about protecting themselves.

How are they to protect themselves? There seems, really, to be only one way, and that is to develop and put into practice the idea of a local economy —something that growing numbers of people are now doing. For several good reasons, they are beginning with the idea of a local food economy. People are trying to find ways to shorten the distance between producers and consumers, to make the connections between the two more direct, and to make this local economic activity a benefit to the local community. They are trying to learn to use the consumer economies of local towns and cities to preserve the livelihoods of local farm families and farm communities. They want to use the local economy to give consumers an influence over the kind and quality of their food, and to preserve and enhance the local landscapes. They want to give everybody in the local community a direct, long-term interest in the prosperity, health, and beauty of their homeland. This is the only way presently available to make the total economy less total. It was once, I believe, the only way to make a national or a colonial economy less total. But now the necessity is greater.

I am assuming that there is a valid line of thought leading from the idea of the total economy to the idea of a local economy. I assume that the first thought may be a recognition of one's ignorance and vulnerability as a consumer in the total economy. As such a consumer, one does not know the history of the products that one uses. Where, exactly, did they come from? Who produced them? What toxins were used in their production? What were the human and ecological costs of producing them and then of disposing of them? One sees that such questions cannot be answered easily, and perhaps not at all. Though one is shopping amid an astonishing variety of products, one is denied certain significant choices. In such a state of economic ignorance it is not possible to choose products that were produced locally or with reasonable kindness toward people and toward nature. Nor is it possible for such consumers to influence production for the better. Consumers who feel a prompting toward land stewardship find that in this economy they can have no stewardly practice. To be a consumer in the total economy, one must agree to be totally ignorant, totally passive, and totally dependent on distant supplies and self-interested suppliers.

And then, perhaps, one begins to *see* from a local point of view. One begins to ask, What is here, what is in me, that can lead to something better? From a local point of view, one can see that a global "free market" economy is possible only if nations and localities accept or ignore the inherent instability of a production economy based on exports and a consumer economy based on imports. An export economy is beyond local influence, and so is an import economy.

And cheap long-distance transport is possible only if granted cheap fuel, international peace, control of terrorism, prevention of sabotage, and the solvency of the international economy.

Perhaps one also begins to see the difference between a small local business that must share the fate of the local community and a large absentee corporation that is set up to escape the fate of the local community by ruining the local community.

* * *

So far as I can see, the idea of a local economy rests upon only two principles: neighborhood and subsistence.

In a viable neighborhood, neighbors ask themselves what they can do or provide for one another, and they find answers that they and their place can afford. This, and nothing else, is the *practice* of neighborhood. This practice must be, in part, charitable, but it must also be economic, and the economic part must be equitable; there is a significant charity in just prices.

Of course, everything needed locally cannot be produced locally. But a viable neighborhood is a community; and a viable community is made up of neighbors who cherish and protect what they have in common. This is the principle of subsistence. A viable community, like a viable farm, protects its own production capacities. It does not import products that it can produce for itself. And it does not export local products until local needs have been met. The economic products of a viable community are understood either as belonging to the community's subsistence or as surplus, and only the surplus is considered to be marketable abroad. A community, if it is to be viable, cannot think of producing solely for export, and it cannot permit importers to use cheaper labor and goods from other places to destroy the local capacity to produce goods that are needed locally. In charity, moreover, it must refuse to import goods that are produced at the cost of human or ecological degradation elsewhere. This principle applies not just to localities, but to regions and nations as well.

The principles of neighborhood and subsistence will be disparaged by the globalists as "protectionism"—and that is exactly what it is. It is a protectionism that is just and sound, because it protects local producers and is the best assurance of adequate supplies to local consumers. And the idea that local needs should be met first and only surpluses exported does not imply any prejudice against charity toward people in other places or trade with them. The principle of neighborhood at home always implies the principle of charity abroad. And the principle of subsistence is in fact the best guarantee of giveable or marketable surpluses. This kind of protection is not "isolationism."

Albert Schweitzer, who knew well the economic situation in the colonies of Africa, wrote nearly sixty years ago: "Whenever the timber trade is good, permanent famine reigns in the Ogowe region because the villagers abandon their farms to fell as many trees as possible." We should notice especially that the goal of production was "as many . . . as possible." And Schweitzer makes my point exactly: "These people could achieve true wealth if they could develop their agriculture and trade to meet their own needs." Instead they produced timber for export to "the world economy," which made them dependent upon imported goods that they bought with money earned from their exports. They gave up their local means of subsistence, and imposed the false standard of a foreign demand ("as many trees as possible") upon their forests. They thus became helplessly dependent on an economy over which they had no control.

Such was the fate of the native people under the African colonialism of Schweitzer's time. Such is, and can only be, the fate of everybody under the global colonialism of our time. Schweitzer's description of the colonial economy of the Ogowe region is in principle not different from the rural economy now in Kentucky or Iowa or Wyoming. A total economy for all practical purposes is a total government. The "free trade," which from the standpoint of the corporate economy brings "unprecedeted economic growth," from the standpoint of the land and its local populations, and ultimately from the standpoint of the cities, is destruction and slavery. Without prosperous local economies, the people have no power and the land no voice.

A Bad Big Idea

After World War II, the United States and seventeen other nations entered into the General Agreement on Tariffs and Trade (also known as GATT) for the purpose of regulating international trade and resolving international trade disputes. Beginning in 1986, with the so-called Uruguay round of GATT negotiations, the Reagan and Bush administrations, working mostly in secret, undertook to make a set of changes in GATT that would have dire economic and ecological effects on the more than one hundred nations now subscribing to the agreement—and that would significantly reduce the freedom of their citizens as well. Whether or not the Clinton administration will continue the Reagan-Bush agitation for these changes remains to be seen.

The U.S. proposals on agriculture were drafted mostly by Daniel Amstutz, formerly a Cargill executive, and they are backed by other large supranational corporations. Made to order for the grain traders and agrochemical companies that operate in the "global economy," these proposals aim both to eliminate farm price supports and production controls and to attempt to force all member nations to conform to health and safety standards that would be set in Rome by Codex Alimentarius, a group of international scientific bureaucrats that is under the influence of the agribusiness corporations. Pressure for these revisions has come solely from these corporations and their allies. There certainly has been no popular movement in favor of them—not in any country—although there have been some popular movements in opposition.

When very important persons have plunder in mind, they characteristically invent ugly euphemisms for what they intend to do, and the promoters of these GATT revisions are no exception:

Tariffication refers to the recommended process by which all controls on imports of agricultural products will be replaced by tariffs, which will then be reduced or eliminated within five to ten years. This would have the effect of opening U.S. markets (and all others) to unlimited imports.

Harmonization refers to a process by which the standards of trade among the member nations would be brought into "harmony." This would mean lowering all those standards regulating food safety, toxic residues, inspections, packaging and labeling, and so on that are higher than the standards set by Codex Alimentarius.

And *fast track* refers to a capitulation by which our Congress has ceded to the president the authority to make an international trade agreement and to draft the enabling legislation, which then is not subject to congressional amendment and which must be accepted or rejected as a whole within ninety session days.

If the proposed revisions in the GATT are adopted, every farmer in every member nation will be thrown into competition with every other farmer. With restrictions lowered to international minimums and with farmers under increasing pressure to make up in volume for drastically reduced unit prices, this will become a competition in land exploitation. Such conservation practices as are now in use (and they are already inadequate) will of necessity be abandoned; land rape and the use of toxic chemicals will increase, as will the exploitation of people. American farmers, who must continue to buy their expensive labor-replacing machines, fuel, and chemicals on markets entirely controlled by the suppliers, will be forced to market their products in competition with the cheapest hand labor of the poor countries. And the poor countries, needing to feed their own people, will see the food vacuumed off their plates by lucrative export markets. The supranational corporations, meanwhile, will be able to slide about at will over the face of the globe to wherever products can be bought cheapest and sold highest.

It is easy to see who will have the freedom in this international "free market." The proposed GATT revisions, as one of their advocates has said, are "exactly what exporters need"—the assumption being, as usual, that what is good for exporters is good for everybody. But what is good for exporters is by no means necessarily good for producers, and in fact these proposed revisions expose a long-standing difference of interest between farmers and agribusiness marketers. We in the United States have seen how unrestrained competition among farmers, increasing surpluses and driving down prices, has directly served the purposes of the agribusiness corporations. These corporations have, in fact, remained hugely and consistently profitable right through an era of severe economic hardship in rural America. They are clearly in a position to take excellent advantage of "free-market" competition, for the proposed GATT revisions would permit them to practice the same exploitation without restraint in the world at large.

What these proposals actually propose is a revolution as audacious, far-reaching, and sudden as any the world has seen. Though they would deny to the people of some 108 nations any choice in the matter of protecting their land, their farmers, their food supply, or their health, these proposals were not drafted and, if adopted, would not be implemented by anybody elected by the people of any of the 108 nations. Their purpose is to bypass all local, state, and national governments in order to subordinate the interests of those governments and of the people they represent to the interests of a global "free market" run by a few supranational corporations. By this single device, if it should be implemented, these corporations would destroy the protections that have been won by generations of conservationists, labor organizers, consumer advocates—and by democrats and lovers of freedom. This is an unabashed attempt to replace government with economics and to destroy any sort of local (let alone personal) self-determination. The intended effect would be to centralize control of all prices and standards in the international food economy and to place this control in the hands of the corporations that are best able to profit from it. The revised GATT would thus be a license issued to a privileged few for an all-out economic assault on the lands and peoples of the world. It would establish a "free" global economy that would be a tighter enclosure than most Americans, at least, have so far experienced.

The issue here really is not whether international trade shall be free but whether or not it makes sense for a country—or, for that matter, a region—to destroy its own capacity to produce its own food. How can a government, entrusted with the safety and health of its people, conscientiously barter away in the name of an economic idea that people's ability to feed itself? And if people lose their ability to feed themselves, how can they be said to be free?

The supporters of these GATT revisions assume that there is no longer any possibility of escape from the global economy and, furthermore, that there is no need for such an escape. They assume that all nations are therefore already properly subservient to the global economy and that the highest purpose of national governments is to serve as attorneys for the supranational corporations. They assume also (like far too many farmers and consumers) that there is no possibility of a food economy that is not decided on "at the top" in some center of power.

But in so assuming, these people unwittingly have provided the rest of us with our best occasion so far to understand and to talk about the need for sound and reasonably self-sufficient local food economies. They have forced us to realize that politics and economics are in fact as inseparable as are economics and ecology. They have made it clear that if we want to be free, we will have to free ourselves somehow from the purposes of these great supranational concentrations of greed, wealth, and power. They have forced us to realize that a General Agreement on Tariffs and Trade may be able to set the standards for governments but that it cannot set the standards for individuals and local communities—unless those individuals and communities allow it to do so. They have, in other words, made certain truths self-evident.

The proposed GATT revisions offend against democracy and freedom, against people's natural concern for bodily and ecological health, and against the very possibility of a sustainable food supply. Apart from the corporate ambition to gather the wealth and power of the world into fewer and fewer hands, these revisions make no sense, for they ignore or reduce to fantasy all the realities with which they are concerned: ecological, agricultural, economic, political, and cultural. Their great evil originates in their underlying assumption that all the world may safely be subjected to the desires and controls of a centralizing power. For this is what "harmonization" really envisions: not the necessary small local harmonies that actually can be made among neighbors and between people and their land but rather the "harmony" that might exist between exploiter and exploited after all protest is silenced and all restraints abandoned. The would-be exploiters of the world would like to assume—it would be so easy for them if they could assume—that the world is everywhere uniform and conformable to their desires.

The world, on the contrary, is made up of an immense diversity of countries, climates, topographies, regions, ecosystems, soils, and human cultures—so many as to be endlessly frustrating to centralizing ambition, and this perhaps explains the attempt to impose a legal uniformity on it. However, anybody who is interested in real harmony, in economic and ecological justice, will see immediately that such justice requires not international uniformity but international generosity toward local diversity.

And anybody interested in solving, rather than profiting from, the problems of food production and distribution will see that in the long run the safest food supply is a local food supply, not a supply that is dependent on a global economy. Nations and regions within nations must be left free — and should be encouraged — to develop the local food economies that best suit local needs and local conditions.

1993

Solving for Pattern (1980)

Our dilemma in agriculture now is that the industrial methods that have so spectacularly solved some of the problems of food production have been accompanied by "side effects" so damaging as to threaten the survival of farming. Perhaps the best clue to the nature and the gravity of this dilemma is that it is not limited to agriculture. My immediate concern here is with the irony of agricultural methods that destroy, first, the health of the soil and, finally, the health of human communities. But I could just as easily be talking about sanitation systems that pollute, school systems that graduate illiterate students, medical cures that cause disease, or nuclear armaments that explode in the midst of the people they are meant to protect. This is a kind of surprise that is characteristic of our time: the cure proves incurable; security results in the evacuation of a neighborhood or a town. It is only when it is understood that our agricultural dilemma is characteristic not of our agriculture but of our time that we can begin to understand why these surprises happen, and to work out standards of judgment that may prevent them.

To the problems of farming, then, as to other problems of our time, there appear to be three kinds of solutions:

There is, first, the solution that causes a ramifying series of new problems, the only limiting criterion being, apparently, that the new problems should arise beyond the purview of the expertise that produced the solution—as, in agriculture, industrial solutions to the problem of production have invariably caused problems of maintenance, conservation, economics, community health, etc., etc.

If, for example, beef cattle are fed in large feed lots, within the boundaries of the feeding operation itself a certain factory-like order and efficiency can be achieved. But even within those boundaries that mechanical order immediately produces biological disorder, for we know that health problems and dependence on drugs will be greater among cattle so confined than among cattle on pasture.

And beyond those boundaries, the problems multiply. Pen feeding of cattle in large numbers involves, first, a manure-removal problem, which becomes at some point a health problem for the animals themselves, for the local watershed, and for the adjoining ecosystems and human communities. If the manure is disposed of without returning it to the soil that produced the feed, a serious problem of soil fertility is involved. But we know too that large concentrations of animals in feed lots in one place tend to be associated with, and to promote, large cash-grain monocultures in other places. These monocultures tend to be accompanied by a whole set of specifically agricultural problems: soil erosion, soil compaction, epidemic infestations of pests, weeds, and disease. But they are also accompanied by a set of agricultural-economic problems (dependence on purchased technology; dependence on purchased fuels, fertilizers, and poisons; dependence on credit)—and by a set of community problems, beginning with depopulation and the removal of sources, services, and markets to more and more distant towns. And these are, so to speak, only the first circle of the bad effects of a bad solution. With a little care, their branchings can be traced on into nature, into the life of the cities, and into the cultural and economic life of the nation.

The second kind of solution is that which immediately worsens the problem it is intended to solve, causing a hellish symbiosis in which problem and solution reciprocally enlarge each other in a sequence that, so far as its own logic is concerned, is limitless—as when the problem of soil compaction is "solved" by a bigger tractor, which further compacts the soil, which makes a need for a still bigger tractor, and so on and on. There is an identical symbiosis between coal-fired power plants and air conditioners. It is characteristic of such solutions that no one prospers by them but the suppliers of fuel and equipment.

These two kinds of solutions are obviously bad. They always serve one good at the expense of another or of several others, and I believe that if all their effects were ever to be accounted for they would be seen to involve, too frequently if not invariably, a net loss to nature, agriculture, and the human commonwealth.

Such solutions always involve a definition of the problem that is either false or so narrow as to be virtually false. To define an agricultural problem as if it were solely a problem of agriculture—or solely a problem of production or technology or economics—is simply to misunderstand the problem, either inadvertently or deliberately, either for profit or because of a prevalent fashion of thought. The whole problem must be solved, not just some handily identifiable and simplifiable aspect of it.

Both kinds of bad solutions leave their problems unsolved. Bigger tractors do not solve the problem of soil compaction any more than air conditioners solve the problem of air pollution. Nor does the large confinement-feeding operation solve the problem of food production; it is, rather, a way calculated to allow large-scale ambition and greed to profit from food production. The real problem of food production occurs within a complex, mutually influential relationship of soil, plants, animals, and people. A real solution to that problem will therefore be ecologically, agriculturally, and culturally healthful.

Perhaps it is not until health is set down as the aim that we come in sight of the third kind of solution: that which causes a ramifying series of solutions—as when meat animals are fed on the farm where the feed is raised, and where the feed is raised to be fed to the animals that are on the farm. Even so rudimentary a description implies a concern for pattern, for quality, which necessarily complicates the concern for production. The farmer has put plants and animals into a relationship of mutual dependence, and must perforce be concerned for balance or symmetry, a reciprocating connection in the pattern of the farm that is biological, not industrial, and that involves solutions to problems of fertility, soil husbandry, economics, sanitation—the whole complex of problems whose proper solutions add up to *health:* the health of the soil, of plants and animals, of farm and farmer, of farm family and farm community, all involved in the same internested, interlocking pattern—or pattern of patterns.

A bad solution is bad, then, because it acts destructively upon the larger patterns in which it is contained. It acts destructively upon those patterns, most likely, because it is formed in ignorance or disregard of them. A bad solution solves for a single purpose or goal, such as increased production. And it is typical of such solutions that they achieve stupendous increases in production at exorbitant biological and social costs.

A good solution is good because it is in harmony with those larger patterns —and this harmony will, I think, be found to have the nature of analogy. A bad solution acts within the larger pattern the way a disease or addiction acts within the body. A good solution acts within the larger pattern the way a healthy organ acts within the body. But it must at once be understood that a healthy organ does not—as the mechanistic or industrial mind would like to say—"give" health to the body, is not exploited for the body's health, but is *a part* of its health. The health of organ and organism is the same, just as the health of organism and ecosystem is the same. And these structures of organ, organism, and ecosystem—as John Todd has so ably understood—belong to a series of analogical integrities that begins with the organelle and ends with the biosphere.

* * *

It would be next to useless, of course, to talk about the possibility of good solutions if none existed in proof and in practice. A part of our work at *The New Farm* has been to locate and understand those farmers whose work is competently responsive to the requirements of health. Representative of these farmers, and among them remarkable for the thoroughness of his intelligence, is Earl F. Spencer, who has a 250-acre dairy farm near Palatine Bridge, New York.

Before 1972, Earl Spencer was following a "conventional" plan that would build his herd to 120 cows. According to this plan, he would eventually buy all the grain he fed, and he was already using as much as 30 tons per year of commercial fertilizer. But in 1972, when he had increased his herd to 70 cows, wet weather reduced his harvest by about half. The choice was clear: he had either to buy half his yearly feed supply, or sell half his herd.

He chose to sell half his herd—a very unconventional choice, which in itself required a lot of independent intelligence. But character and intelligence of an even more respectable order were involved in the next step, which was to understand that the initial decision implied a profound change in the pattern of the farm and of his life and assumptions as a farmer. With his herd now reduced by half, he saw that before the sale he had been overstocked, and had been abusing his land. On his 120 acres of tillable land, he had been growing 60 acres of corn and 60 of alfalfa. On most of his fields, he was growing corn three years in succession. The consequences of this he now saw as symptoms, and saw that they were serious: heavy dependence on purchased supplies, deteriorating soil structure, declining quantities of organic matter, increasing erosion, yield reductions despite continued large applications of fertilizer. In addition, because of his heavy feeding of concentrates, his cows were having serious digestive and other health problems.

He began to ask fundamental questions about the nature of the creatures and the land he was dealing with, and to ask if he could not bring about some sort of balance between their needs and his own. His conclusion was that "to be in balance with nature is to be successful." His farm, he says, had been going in a "dead run"; now he would slow it to a "walk."

From his crucial decision to reduce his herd, then, several other practical measures have followed:

1. A five-year plan (extended to eight years) to phase out entirely his use of purchased fertilizers.
2. A plan, involving construction of a concrete manure pit, to increase and improve his use of manure.
3. Better husbandry of cropland, more frequent rotation, better timing.
4. The gradual reduction of grain in the feed ration, and the concurrent increase of roughage—which has, to date, reduced the dependence on grain by half, from about 6,000 pounds per cow to about 3,000 pounds.
5. A breeding program which selects "for more efficient roughage conversion."

The most tangible results are that the costs of production have been "dramatically" reduced, and that per cow production has increased by 1,500 to 2,000 pounds. But the health of the whole farm has improved. There is a moral satisfaction in this, of which Earl Spencer is fully aware. But he is also aware that the satisfaction is not *purely* moral, for the good results are also practical and economic: "We have half the animals we had before and are feeding half as much grain to those remaining, so we now need to plant corn only two years in a row. Less corn means less plowing, less fuel for growing and harvesting, and less wear on the most expensive equipment." Veterinary bills have been reduced also. And in 1981, if the schedule holds, he will buy no commercial fertilizer at all.

* * *

From the work of Earl Spencer and other exemplary farmers, and from the understanding of destructive farming practices, it is possible to devise a set of critical standards for agriculture. I am aware that the list of standards that follows must be to some extent provisional, but am nevertheless confident that it will work to distinguish between healthy and unhealthy farms, as well as between the oversimplified minds that solve problems for some X such as profit or quantity of production, and those minds, sufficiently complex, that solve for health or quality or coherence of pattern. To me, the validity of these standards seems inherent in their general applicability. They will serve the making of sewer systems or households as readily as they will serve the making of farms:

1. A good solution accepts given limits, using so far as possible what is at hand. The farther-fetched the solution, the less it should be trusted. Granted that a farm can be too small, it is nevertheless true that enlarging scale is a deceptive solution; it solves one problem by acquiring another or several others.

2. A good solution accepts also the limitation of discipline. Agricultural problems should receive solutions that are agricultural, not technological or economic.

3. A good solution improves the balances, symmetries, or harmonies within a pattern—it is a qualitative solution—rather than enlarging or complicating some part of a pattern at the expense or in neglect of the rest.

4. A good solution solves more than one problem, and it does not make new problems. I am talking about health as opposed to almost any cure, coherence of pattern as opposed to almost any solution produced piecemeal or in isolation. The return of organic wastes to the soil may, at first glance, appear to be a good solution *per se*. But that is not invariably or necessarily true. It is true only if the wastes are returned to the right place at the right time in the pattern of the farm, if the waste does not contain toxic materials, if the quantity is not too great, and if not too much energy or money is expended in transporting it.

5. A good solution will satisfy a whole range of criteria; it will be good in all respects. A farm that has found correct agricultural solutions to its problems will be fertile, productive, healthful, conservative, beautiful, pleasant to live on. This standard obviously must be qualified to the extent that the pattern of the life of a farm will be adversely affected by distortions in any of the larger patterns that contain it. It is hard, for instance, for the economy of a farm to maintain its health in a national industrial economy in which farm earnings are apt to be low and expenses high. But it is apparently true, even in such an economy, that the farmers most apt to survive are those who do not go too far out of agriculture into either industry or banking—and who, moreover, live like farmers, not like businessmen. This seems especially true for the smaller farmers.

6. A good solution embodies a clear distinction between biological order

and mechanical order, between farming and industry. Farmers who fail to make this distinction are ideal customers of the equipment companies, but they often fail to understand that the real strength of a farm is in the soil.

7. Good solutions have wide margins, so that the failure of one solution does not imply the impossibility of another. Industrial agriculture tends to put its eggs into fewer and fewer baskets, and to make "going for broke" its only way of going. But to grow grain should not make it impossible to pasture livestock, and to have a lot of power should not make it impossible to use only a little.

8. A good solution always answers the question, How much is enough? Industrial solutions have always rested on the assumption that enough is all you can get. But that destroys agriculture, as it destroys nature and culture. The good health of a farm implies a limit of scale, because it implies a limit of attention, and because such a limit is invariably implied by any pattern. You destroy a square, for example, by enlarging one angle or lengthening one side. And in any sort of work there is a point past which more quantity necessarily implies less quality. In some kinds of industrial agriculture, such as cash grain farming, it is possible (to borrow an insight from Professor Timothy Taylor) to think of technology as a substitute for skill. But even in such farming that possibility is illusory; the illusion can be maintained only so long as the consequences can be ignored. The illusion is much shorter lived when animals are included in the farm pattern, because the husbandry of animals is so insistently a human skill. A healthy farm incorporates a pattern that a single human mind can comprehend, make, maintain, vary in response to circumstances, and pay steady attention to. That this limit is obviously variable from one farmer and farm to another does not mean that it does not exist.

9. A good solution should be cheap, and it should not enrich one person by the distress or impoverishment of another. In agriculture, so-called "inputs" are, from a different point of view, outputs—*expenses*. In all things, I think, but especially in an agriculture struggling to survive in an industrial economy, any solution that calls for an expenditure to a manufacturer should be held in suspicion—not rejected necessarily, but *as a rule* mistrusted.

10. Good solutions exist only in proof, and are not to be expected from absentee owners or absentee experts. Problems must be solved in work and in place, with particular knowledge, fidelity, and care, by people who will suffer the consequences of their mistakes. There is no theoretical or ideal *practice*. Practical advice or direction from people who have no practice may have some value, but its value is questionable and is limited. The divisions of capi-

tal, management, and labor, characteristic of an industrial system, are therefore utterly alien to the health of farming—as they probably also are to the health of manufacturing. The good health of a farm depends on the farmer's mind; the good health of his mind has its dependence, and its proof, in physical work. The good farmer's mind and his body—his management and his labor—work together as intimately as his heart and his lungs. And the capital of a well-farmed farm by definition includes the farmer, mind and body both. Farmer and farm are one thing, an organism.

11. Once the farmer's mind, his body, and his farm are understood as a single organism, and once it is understood that the question of the endurance of this organism is a question about the sufficiency and integrity of a pattern, then the word *organic* can be usefully admitted into this series of standards. It is a word that I have been defining all along, though I have not used it. An organic farm, properly speaking, is not one that uses certain methods and substances and avoids others; it is a farm whose structure is formed in imitation of the structure of a natural system; it has the integrity, the independence, and the benign dependence of an organism. Sir Albert Howard said that a good farm is an analogue of the forest that "manures itself." A farm that imports too much fertility, even as feed or manure, is in this sense as inorganic as a farm that exports too much or that imports chemical fertilizer.

12. The introduction of the term *organic* permits me to say more plainly and usefully some things that I have said or implied earlier. In an organism, what is good for one part is good for another. What is good for the mind is good for the body; what is good for the arm is good for the heart. We know that sometimes a part may be sacrificed for the whole; a life may be saved by the amputation of an arm. But we also know that such remedies are desperate, irreversible, and destructive; it is impossible to improve the body by amputation. And such remedies do not imply a safe logic. As *tendencies* they are fatal: you cannot save your arm by the sacrifice of your life.

Perhaps most of us who know local histories of agriculture know of fields that in hard times have been sacrificed to save a farm, and we know that though such a thing is possible it is dangerous. The danger is worse when topsoil is sacrificed for the sake of a crop. And if we understand the farm as an organism, we see that it is impossible to sacrifice the health of the soil to improve the health of plants, or to sacrifice the health of plants to improve the health of animals, or to sacrifice the health of animals to improve the health of people. In a biological pattern—as in the pattern of a community—the exploitive means and motives of industrial economics are immediately destructive and ultimately suicidal.

13. It is the nature of any organic pattern to be contained within a larger one. And so a good solution in one pattern preserves the integrity of the pattern that contains it. A good agricultural solution, for example, would not pollute or erode a watershed. What is good for the water is good for the ground, what is good for the ground is good for plants, what is good for plants is good for animals, what is good for animals is good for people, what is good for people is good for the air, what is good for the air is good for the water. And vice versa.

14. But we must not forget that those human solutions that we may call organic are not natural. We are talking about organic *artifacts*, organic only by imitation or analogy. Our ability to make such artifacts depends on virtues that are specifically human: accurate memory, observation, insight, imagination, inventiveness, reverence, devotion, fidelity, restraint. Restraint—for us, now—above all: the ability to accept and live within limits; to resist changes that are merely novel or fashionable; to resist greed and pride; to resist the temptation to "solve" problems by ignoring them, accepting them as "trade-offs," or bequeathing them to posterity. A good solution, then, must be in harmony with good character, cultural value, and moral law.

PART V

Agrarian Religion

To articulate an agrarian ethos is finally to confront questions that are religious, since what is at stake in agrarianism is responsibility for the grace and mystery of life. The history of religions, particularly as they have been guided by other-worldly ambitions, shows that faith communities have not always been the stewards of life. In fact, they have in specific cases been the agents or abettors of the destruction of the earth. In the following essays Berry calls religious communities, especially Christian communities, to a recovery of the sense of life's sanctity. From this sense, a sense grounded in the agrarian emphasis on the gifts of grace, a spirituality of responsibility and conviviality can emerge.

The Use of Energy

"Energy," said William Blake, "is Eternal Delight." And the scientific prognosticators of our time have begun to speak of the eventual opening, for human use, of "infinite" sources of energy. In speaking of the use of energy, then, we are speaking of an issue of religion, whether we like it or not.

Religion, in the root sense of the word, is what binds us back to the source of life. Blake also said that "Energy is the only life . . ." And it is superhuman in the sense that humans cannot create it. They can only refine or convert it. And they are bound to it by one of the paradoxes of religion: they cannot have it except by losing it; they cannot use it except by destroying it. The lives that feed us have to be killed before they enter our mouths; we can only use the fossil fuels by burning them up. We speak of electrical energy as "current": it exists only while it runs away; we use it only by delaying its escape. To receive energy is at once to live and to die.

Perhaps from an "objective" point of view it is incorrect to say that we can destroy energy; we can only change it. Or we can destroy it only in its current form. But from a human point of view, we can destroy it also by wasting it—that is, by changing it into a form in which we cannot use it again. As users, we can preserve energy in cycles of use, passing it again and again through the same series of forms; or we can waste it by using it once in a way that makes it irrecoverable. The human pattern of cyclic use is exemplified in the small Oriental peasant farms described in F. H. King's *Farmers of Forty Centuries*, in which all organic residues, plant and animal and human, were returned to the soil, thus keeping intact the natural cycle of "birth, growth, maturity, death, and decay" that Sir Albert Howard identified as the "Wheel of Life." The pattern of wasteful use is exemplified in the modern sewage system and the internal combustion engine. With us, the wastes that escape use typically become pollutants. This kind of use turns an asset into a liability.

We have two means of bringing energy to use: by living things (plants, animals, our own bodies) and by tools (machines, energy-harnesses). For the use of these we have skills or techniques. All three together comprise our technology. Technology joins us to energy, to life. It is not, as many technologists would have us believe, a simple connection. Our technology is the practical aspect of our culture. By it we enact our religion, or our lack of it.

I began thinking about this by trying to make a clear distinction between the living organisms and skills of technology and its mechanisms, and to say that the living aspect was better than the mechanical. I found it impossible to make such a distinction. I thought of going back through history to a point at which such a distinction would become possible, but found that the farther back I went the less possible it became. When people had no machines other than throwing stones and clubs, their technology was all of a piece. It stayed that way through their development of more sophisticated tools, their mastery of fire, their domestication of plants and animals. Lives, skills, and tools were culturally indivisible.

The question at issue, then, is not of distinction but of balance. The ideal seems to be that the living part of our technology should not be devalued or overpowered by the mechanical. Because the biological limits are probably narrower than the mechanical, this calls for restraint on the proliferation of machines.

At some point in history the balance between life and machinery was overthrown. I think this began to happen when people began to desire long-term stores or supplies of energy—that is, when they began to think of energy as volume as well as force—and when machines ceased to enhance or elaborate skill and began to replace it.

Though it seems impossible to distinguish between the living and the mechanical aspects of technology, it is possible to distinguish between two kinds of energy: that which is made available by living things and that which is made available by machines.

The energy that comes from living things is produced by combining the four elements of medieval science: earth, air, fire (sunlight), and water. This is current energy. Though it is possible to speak of a *reserve* of such energy, as Sir Albert Howard does, in the sense of a surplus of fertility, it is impossible to conceive of a *reservoir* of it. It is not available in long-term supplies; in any form in which it can be preserved, as in humus, in the flesh of living animals, in cans or freezers or grain elevators, it still perishes fairly quickly in comparison, say, to coal or plutonium. It lasts over a long term only in the living cycle of birth, growth, maturity, death, and decay. The technology appropriate to the use of this energy, therefore, preserves its cycles. It is a technology that never escapes into its own logic but remains bound in analogy to natural law.

The energy that is made available, and consumed, by machines is typically energy that can be accumulated in stockpiles or reservoirs. Energy from wind and water obviously does not fit this category, but it suggests the possibility of bigger and better storage batteries, which one must assume will sooner or later be produced. And, of course, we already store water power behind hydroelectric dams. This mechanically derived energy is supposed to have set people free from work and other difficulties once considered native to the human condition. Whether or not it has done so in any meaningful sense is questionable — in my opinion, it is highly questionable. But there is no doubt that this sort of energy has freed machinery from the natural restraints that apply to the use of organic energy. We now have a purely mechanical technology that is very nearly a law unto itself.

And yet, in the long term, this liberation of the machine is illusory. Mechanical technology is based on quantities of materials and fuels that are finite. If the prophets of science foresee "limitless abundance" and "infinite resources," one must assume that they are speaking figuratively, meaning simply that they cannot comprehend how much there may be. In that sense, they are right: there are sources of energy that, given the necessary machinery, are inexhaustible *as far as we can see*.

The great difficulty, which these cheerful prophets do not acknowledge at all, is that we are trustworthy only so far as we can see. The length of our vision is our moral boundary. Even if these foreseen supplies *are* limitless, we can use them only within limits. We can bring the infinite to bear only within the finite bounds of our biological circumstance and our understanding. It is already certain that our planet alone—not to mention potential sources in space—can provide us with more energy and materials than we can use safely or well. By our abuse of our finite sources, our lives and all life are already in danger. What might we bring into danger by the abuse of "infinite" sources?

The difficulty with mechanically extractable energy is that so far we have been unable to make it available without serious geological and ecological damage, or to effectively restrain its use, or to use or even neutralize its wastes. From birth, right now, we are carrying the physical and the moral poisons produced by our crude and ignorant use of this sort of energy. And the more abundant the energy of this sort that we use, the more abounding must be the consequences.

It is typical of the mentality of our age that we cannot conceive of infinity except as an enormous quantity. We cannot conceive of it as orderly process, as pattern or cycle, as shapeliness. We conceive of it as inconceivable quantity—that is, as the immeasurable. Any quantity that we cannot measure we assume must be infinite. That is about as sophisticated as saying that the world is flat because it *looks* flat. The talk about "infinite" resources is thus a kind of scientific-sounding foolishness. And it involves some quaint paradoxes. If we think, for instance, of infinite energy as immeasurable fuel, we are committed in the same thought to its destruction, for fuel must be destroyed to be used. We thus arrive at the curious idea of a destructible infinity. Furthermore, we have become guilty not only of the demonstrably silly assumption that we know what to do with infinite energy, but also of the monstrous pride of thinking ourselves somehow entitled to undertake infinite destruction.

This mechanically rendered infinitude of energy is an ambition surrounded by terrific problems. Such energy cannot be used constructively without at the same time being used destructively. And which way the balance will finally fall is a question that baffles the best minds. Nobody knows what will be the ultimate consequences of our present use of fossil fuel, much less those of our future use of atomic fuel. The sun may prove an "infinite" source of energy—at least one that may last several billion years. But who will control the use of that energy? How and for what purposes will it be used? How much can be used without overthrowing ecological or social or political balances? Nobody knows.

The energy that is made available to us by living things, on the other hand, is made available not as an inconceivable quantity, but as a conceivable pattern. And for the mastery of this pattern—that is, the ability to see its absolute importance and to preserve it in use—one does not need a Ph.D. or a laboratory or a computer. One can master it in this sense, in fact, without having any analytic or scientific understanding of it at all. It was mastered, better than our scientific experts have mastered it, by "primitive" peasants and tribesmen thousands of years before modern science. It is conceivable not so much to the analytic intelligence, to which it may always remain in part mysterious, as to the imagination, by which we perceive, value, and imitate order beyond our understanding.

We cannot create biological energy any more than we can create atomic or fossil fuel energy. But we *can* preserve it in use; we can probably even augment it in use, in the sense that, by proper care, we can "build" soil. We cannot do that with machine-derived energy. This is an extremely important difference, with respect both to the energy economy itself and to the moral order that is undoubtedly determined by, as much as it determines, the value we put on energy.

The moral order by which we use machine-derived energy is comparatively simple. Whatever uses this sort of energy works simply as a conduit that carries it beyond use: the energy goes in as "fuel" and comes out as "waste." This principle sustains a highly simplified economy having only two functions: production and consumption.

The moral order appropriate to the use of biological energy, on the other hand, requires the addition of a third term: production, consumption, *and return*. It is the principle of return that complicates matters, for it requires responsibility, care, of a different and higher order than that required by production and consumption alone, and it calls for methods and economies of a different kind. In an energy economy appropriate to the use of biological energy, all bodies, plant and animal and human, are joined in a kind of energy community. They are not divided from each other by greedy, "individualistic" efforts to produce and consume large quantities of energy, much less to store large quantities of it. They are indissolubly linked in complex patterns of energy exchange. They die into each other's life, live into each other's death. They do not consume in the sense of using up. They do not produce waste. What they take in they change, but they change it always into a form necessary for its use by a living body of another kind. And this exchange goes on and on, round and round, the Wheel of Life rising out of the soil, descending into it, through the bodies of creatures.

The soil is the great connector of lives, the source and destination of all. It is the healer and restorer and resurrector, by which disease passes into health, age into youth, death into life. Without proper care for it we can have no community, because without proper care for it we can have no life.

It is alive itself. It is a grave, too, of course. Or a healthy soil is. It is full of dead animals and plants, bodies that have passed through other bodies. For except for some humans — with their sealed coffins and vaults, their pathological fear of the earth — the only way into the soil is through other bodies. But no matter how finely the dead are broken down, or how many times they are eaten, they yet give into other life. If a healthy soil is full of death it is also full of life: worms, fungi, microorganisms of all kinds, for which, as for us humans, the dead bodies of the once living are a feast. Eventually this dead matter becomes soluble, available as food for plants, and life begins to rise up again, out of the soil into the light. Given only the health of the soil, nothing that dies is dead for very long. Within this powerful economy, it seems that death occurs only for the good of life. And having followed the cycle around, we see that we have not only a description of the fundamental biological process, but also a metaphor of great beauty and power. It is impossible to contemplate the life of the soil for very long without seeing it as analogous to the life of the spirit. No less than the faithful of religion is the good farmer mindful of the persistence of life through death, the passage of energy through changing forms.

And this living topsoil — living in both the biological sense and in the cultural sense, as metaphor — is the basic element in the technology of farming.

It is the nature of the soil to be highly complex and variable, to conform very inexactly to human conclusions and rules. It is itself a pattern of inexhaustible intricacy, and so it is easily damaged by the imposition of alien patterns. Out of the random grammar and lexicon of possibilities — geological, topographical, climatological, biological — the soil of any one place makes its own peculiar and inevitable sense. It makes an order, a pattern of forms, kinds, and processes, that includes any number of offsets and variables. By its permeability and absorbency, for example, the healthy soil corrects the irregularities of rainfall; by the diversity of its vegetation it protects against both disease and erosion. Most farms, even most fields, are made up of different kinds of soil patterns or soil sense. Good farmers have always known this and have used the land accordingly; they have been careful students of the natural vegetation, soil depth and structure, slope and drainage. They are not appliers of generalizations, theoretical or methodological or mechanical. Nor are they the active agents of their own economic will, working their way upon an inert and passive mass. They are responsive partners in an intimate and mutual relationship.

Because the soil is alive, various, intricate, and because its processes yield more readily to imitation than to analysis, more readily to care than to coercion, agriculture can never be an exact science. There is an inescapable kinship between farming and art, for farming depends as much on character, devotion, imagination, and the sense of structure, as on knowledge. It is a practical art.

But it is also a practical religion, a practice of religion, a rite. By farming we enact our fundamental connection with energy and matter, light and darkness. In the cycles of farming, which carry the elemental energy again and again through the seasons and the bodies of living things, we recognize the only infinitude within reach of the imagination. How long this cycling of energy will continue we do not know; it will have to end, at least here on this planet, sometime within the remaining life of the sun. But by aligning ourselves with it here, in our little time within the unimaginable time of the sun's burning, we touch infinity; we align ourselves with the universal law that brought the cycles into being and that will survive them.

The word *agriculture*, after all, does not mean "agriscience," much less "agribusiness." It means "cultivation of land." And *cultivation* is at the root of the sense both of *culture* and of *cult*. The ideas of tillage and worship are thus joined in *culture*. And these words all come from an Indo-European root meaning both "to revolve" and "to dwell." To live, to survive on the earth, to care for the soil, and to worship, all are bound at the root to the idea of a cycle. It is only by understanding the cultural complexity and largeness of the concept of agriculture that we can see the threatening diminishments implied by the term "agribusiness."

That agriculture is in so complex a sense a cultural endeavor—and that food is therefore a cultural product—would be regarded as heresy by most of the agencies, institutions, and publications of modern farming. The spokesmen of the official reckoning would doubtless respond that they are not cultural but scientific, that they are specialists of "agriscience." If agriculture is acknowledged to have anything to do with culture, then its study has to include people. But the agriculture experts ruled people out when they made their discipline a specialty—or, rather, when they sorted it into a collection of specialties—and moved it into its own "college" in the university. This specialty collection is interested in soils (in the limited sense of soil chemistry), in plants and animals, and in machines and chemicals. It is not interested in people.

But what respect is one to give to a science that parcels a unified discipline into discrete fragments, that has no interest in its effects if they are not immediately measurable in a laboratory, and that is founded upon the waste of topsoil, energy, and manpower, and upon the dissolution of communities? Not much. And it has been my experience that, with respect to this science, farmers are divided into two kinds: those who endanger their solvency, and often their sanity, by trusting it and those who hold it in contempt.

In the view of the experts, then, agriculture is not only not a concern of culture, but not even a concern of science, for they have abandoned interest in the health of the farming communities on the one hand and in the health of the land on the other. They appear to have concluded that agriculture is purely a commercial concern; its purpose is to provide as much food as quickly and cheaply and with as few man-hours as possible and to be a market for machines and chemicals. It is, after all, "agribusiness"—not the land or the farming people—that now benefits most from agricultural research and that can promote humble academicians to highly remunerative and powerful positions in corporations and in government. Former Secretary Earl Butz's career exemplifies the predominant direction of interest of the agriculture specialist. According to Lauren Soth, writing in the *Nation*, "Butz is the perfect example of the agribusiness, commercial-farming, agricultural-education establishment man. When dean of agriculture at Purdue University, he also sat on the boards of directors of the Ralston-Purina Co., the J. I. Case Co., International Minerals and Chemicals Corp., Stokely-Van Camp Co. and Standard Life Insurance Co. of Indiana." By such men and such careers the land-grant college system, originally meant to enhance the small-farm possibility, has been captured for the corporations.

The discipline of agriculture—the "great subject," as Sir Albert Howard called it, "of health in soil, plant, animal, and man"—has been reduced to fit first the views of a piecemeal "science" and then the purposes of corporate commerce. I can see no possibility of a doubt that this is true, though I cannot explain exactly how it happened. But it seems to me that the way was prepared when the specialized shapers or makers of agricultural thought simplified their understanding of energy and began to treat current, living, biological energy as if it were a *store* of energy extractable by machinery. At that point the living part of technology began to be overpowered by the mechanical. The machine was on its own, to follow its own logic of elaboration and growth apart from life, the standard that had previously defined its purposes and hence its limits. Let loose from any moral standard or limit, the machine was also let loose in another way: it replaced the Wheel of Life as the governing cultural metaphor. Life came to be seen as a road, to be traveled as fast as possible, never to return. Or, to put it another way, the Wheel of Life became an industrial metaphor; rather than turning in place, revolving in order to dwell, it began to roll on the "highway of progress" toward an ever-receding horizon. The idea, the responsibility, of return weakened and disappeared from agricultural discipline. Henceforth, *any* resource would be regarded as an ore.

If agriculture is founded upon life, upon the use of living energy to serve human life, and if its primary purpose must therefore be to preserve the integrity of the life cycle, then agricultural technology must be bound under the rule of life. It must conform to natural processes and limits rather than to mechanical or economic models. The culture that sustains agriculture and that it sustains must form its consciousness and its aspiration upon the correct metaphor of the Wheel of Life. The appropriate agricultural technology would therefore be diverse; it would aspire to diversity; it would enable the diversification of economies, methods, and species to conform to the diverse kinds of land. It would always use plants and animals together. It would be as attentive to decay as to growth, to maintenance as to production. It would return all wastes to the soil, control erosion, and conserve water. To enable care and devotion and to safeguard the local communities and cultures of agriculture, it would use the land in small holdings. It would aspire to make each farm so far as possible the source of its own operating energy, by the use of human energy, work animals, methane, wind or water or solar power. The mechanical aspect of the technology would serve to harness or enhance the energy available on the farm. It would not be permitted to replace such energies with imported fuels, to replace people, or to replace or reduce human skills.

The damages of our present agriculture all come from the determination to use the life of the soil as if it were an extractable resource like coal, to use living things as if they were machines, to impose scientific (that is, laboratory) exactitude upon living complexities that are ultimately mysterious.

If animals are regarded as machines, they are confined in pens remote from the source of their food, where their excrement becomes, instead of a fertilizer, first a "waste" and then a pollutant. Furthermore, because confinement feeding depends so largely on grains, grass is removed from the rotation of crops and more land is exposed to erosion.

If plants are regarded as machines, we wind up with huge monocultures, productive of elaborate ecological mischiefs, which are in turn productive of agricultural mischief: monocultures are much more susceptible to pests and diseases than mixed cultures and are therefore more dependent on chemicals.

If the soil is regarded as a machine, then its life, its involvement in living systems and cycles, must perforce be ignored. It must be treated as a dead, inert chemical mass. If its life is ignored, then so must be the natural sources of its fertility—and not only ignored, but scorned. Alfalfa and the clovers, according to some of the most up-to-date practitioners, are "weeds"; the only legitimate source of nitrogen is the fertilizer manufacturer. And animal manures are "wastes"; "efficiency" cannot use them. Not long ago I found that the manure from a saddle-horse barn belonging to the University of Kentucky was simply being dumped. When I asked why it was not used somewhere on the farm, I was told that it would interfere with the College of Agriculture's experiments. The result is absurd: our agriculture, potentially capable of a large measure of independence, is absolutely dependent on petroleum, on the oil companies, and on the vagaries of politics.

If people are regarded as machines, they must be regarded as replaceable by other machines. They are regarded, in other words, as dispensable. Their place on the farm is safe only as long as they are mechanically necessary.

In modern agriculture, then, the machine metaphor is allowed to usurp and wipe from consideration not merely *some* values, but the very *issue* of value. Once the expert's interest is focused on the question of "what will work" within the exclusive confines of his theoretical model, values are no longer of any concern whatever. The confines of his specialty enable him to impose a biological totalitarianism on—he thinks, since he is an agricultural expert—the farm. When he leaves his office or laboratory he will, he assumes, go "home" to value.

But then it must be asked if we can remove cultural value from one part of our lives without destroying it also in the other parts. Can we justify secrecy, lying, and burglary in our so-called intelligence organizations and yet preserve openness, honesty, and devotion to principle in the rest of our government? Can we subsidize mayhem in the military establishment and yet have peace, order, and respect for human life in the city streets? Can we degrade all forms of essential work and yet expect arts and graces to flourish on weekends? And can we ignore all questions of value on the farm and yet have them answered affirmatively in the grocery store and the household?

The answer is that, though such distinctions can be made theoretically, they cannot be preserved in practice. Values may be corrupted or abolished in only one discipline at the start, but the damage must sooner or later spread to all; it can no more be confined than air pollution. If we corrupt agriculture we corrupt culture, for in nature and within certain invariable social necessities we are one body, and what afflicts the hand will afflict the brain.

The effective knowledge of this unity must reside not so much in doctrine as in skill. Skill, in the best sense, is the enactment or the acknowledgment or the signature of responsibility to other lives; it is the practical understanding of value. Its opposite is not merely unskillfulness, but ignorance of sources, dependences, relationships.

Skill is the connection between life and tools, or life and machines. Once, skill was defined ultimately in qualitative terms: How *well* did a person work; how good, durable, and pleasing were his products? But as machines have grown larger and more complex, and as our awe of them and our desire for labor-saving have grown, we have tended more and more to define skill quantitatively: How speedily and cheaply can a person work? We have increasingly wanted a *measurable* skill. And the more quantifiable skills became, the easier they were to replace with machines. As machines replace skill, they disconnect themselves from life; they come between us and life. They begin to enact our ignorance of value—of essential sources, dependences, and relationships.

The catch is that we cannot live in machines. We can only live in the world, in life. To live, our contact with the sources of life must remain direct: we must eat, drink, breathe, move, mate, etc. When we let machines and machine skills obscure the values that represent these fundamental dependences, then we inevitably damage the world; we diminish life. We begin to "prosper" at the cost of a fundamental degradation.

The digging stick, for example, brought in a profound technological revolution: it made agriculture possible. Its use required skill. But its *effect* also required skill, and this kind of skill was higher and more complex than the first, for it involved restraint and responsibility. The digging stick made it possible to grow food; that was one thing. It also made it possible, and necessary, to disturb the earth; and that was another thing. The first skill required others that were its moral elaboration: the skill used in disturbing the earth called directly for other skills that would preserve the earth and restore its fertility.

Until fairly recently, as agricultural tools became more efficient or powerful or both, they required an increase of both kinds of skill. One could do more with stone implements than with sticks, and more with metal implements than with stone implements; the skilled use of these tools enabled one to disturb more ground and so called for further elaboration of the skills of responsibility.

This remained true after the beginning of the use of draft animals. The skills of use had to become much greater, for the human mind had to relate to the animal mind in a new way: not by the magic and cunning of the hunt, but in the practical intricacies of collaboration. And the skills of responsibility had to increase proportionately. More ground could now be disturbed, and so the technology of preservation had to become much larger. Also, the investment of life in work greatly increased; people had to take responsibility not only for their own appetites and excrements but for those of their animals as well.

It was only with the introduction of self-powering machines, and of machine-extracted energy, into the fields that something really new happened to agricultural skills: they began a radical diminishment.

In the first place, it requires more skill to use a team of horses or mules or oxen than to use a tractor. It is more difficult to learn to manage an animal than a machine; it takes longer. Two minds and two wills are involved. A relationship between a person and a work animal is analogous to a relationship between two people. Success depends upon the animal's willingness and upon its health; certain moral imperatives and restraints are therefore pragmatically essential. No such relationship is either necessary or possible with a machine. Within the range of the possible, a machine is directly responsive to human will; it neither starts nor stops because it wants to. A machine has no life, and for this reason it cannot of itself impose any restraint or any moral limit on behavior.

In the second place, the substitution of machines for work animals is justified mainly by their ability to increase the volume of work per man—that is, by their greater speed. But as speed increases, care declines. And so, necessarily, do the skills of responsibility. If this were not so, we would not restrict the speed of traffic in residential areas. We know that there is a limit to the capacity of attention, and that the faster we go the less we see. This law applies with equal force to work; the faster we work the less attention we can pay to its details, and the less skill we can apply to it.

This is true of *any* productive work, and it has great cultural importance; at present we are all suffering, in various ways, from dependence on goods that are poorly made. But its importance in agricultural production is probably more critical than elsewhere. In any biological system the first principle is restraint—that is, the natural or moral checks that maintain a balance between use and continuity. The life of one year must not be allowed to diminish the life of the next; nothing must live at the expense of the source. Thus, in nature, the food species is dependent on its predator, and pests and diseases are agents of health; so populations are controlled and balanced. In agriculture these natural checks are removed and therefore must be replaced by the skills of responsibility, which have to do with the prevention of erosion, the diversification and rotation of plant and animal species, the return of wastes to the soil, and all the other provisionings of the source. When productive power—that is, speed—in machines replaces the productive skills of people, there is a consequent narrowing of attention. The machines are expensive and they run on purchased fuels; they feed upon money. The work of production is immediately profitable, whereas the work of responsibility is not. Once the machine is in the field it creates an economic pressure that enforces haste; the machine concentrates all the energy of the farm and hurries it toward the marketplace. The demands of immediate use eclipse the demands of continuity. As the skills of production decline, the skills of responsibility perish.

To argue for a balance between people and their tools, between life and machinery, between biological and machine-produced energy, is to argue for restraint upon the use of machines. The arguments that rise out of the machine metaphor—arguments for cheapness, efficiency, labor-saving, economic growth, etc.—all point to infinite industrial growth and infinite energy consumption. The moral argument points to restraint; it is a conclusion that may be in some sense tragic, but there is no escaping it. Much as we long for infinities of power and duration, we have no evidence that these lie within our reach, much less within our responsibility. It is more likely that we will have either to live within our limits, within the human definition, or not live at all. And certainly the knowledge of these limits and of how to live within them is the most comely and graceful knowledge that we have, the most healing and the most whole.

The knowledge that purports to be leading us to transcendence of our limits has been with us a long time. It thrives by offering material means of fulfilling a spiritual, and therefore materially unappeasable, craving: we would all very much like to be immortal, infallible, free of doubt, at rest. It is because this need is so large, and so different in kind from all material means, that the knowledge of transcendence—our entire history of scientific "miracles"—is so tentative, fragmentary, and grotesque. Though there are undoubtedly mechanical limits, because there are human limits, there is no mechanical restraint. The only logic of the machine is to get bigger and more elaborate. In the absence of *moral* restraint—and we have never imposed adequate moral restraint upon our use of machines—the machine is out of control by definition. From the beginning of the history of machine-developed energy, we have been able to harness more power than we could use responsibly. From the beginning, these machines have created effects that society could absorb only at the cost of suffering and disorder.

And so the issue is not of supply but of use. The energy crisis is not a crisis of technology but of morality. We already have available more power than we have so far dared to use. If, like the strip miners and the "agribusinessmen," we look on all the world as fuel or as extractable energy, we can do nothing but destroy it. The issue is restraint. The energy crisis reduces to a single question: Can we forbear to do anything that we are able to do? Or to put the question in the words of Ivan Illich: Can we, believing in "the effectiveness of power," see "the disproportionately greater effectiveness of abstaining from its use"?

The only people among us that I know of who have answered this question convincingly in the affirmative are the Amish. They alone, as a community, have carefully restricted their use of machine-developed energy, and so have become the only true masters of technology. They are mostly farmers, and they do most of their farmwork by hand and by the use of horses and mules. They are pacifists, they operate their own local schools, and in other ways hold themselves aloof from the ambitions of a machine-based society. And by doing so they have maintained the integrity of their families, their community, their religion, and their way of life. They have escaped the mainstream American life of distraction, haste, aimlessness, violence, and disintegration. Their life is not idly wasteful, or destructive. The Amish no doubt have their problems; I do not wish to imply that they are perfect. But it cannot be denied that they have mastered one of the fundamental paradoxes of our condition: we can make ourselves whole only by accepting our partiality, by living within our limits, by being human—not by trying to be gods. By restraint they make themselves whole.

"Dream not of other Worlds . . . "

Paradise Lost VIII, 175

The Gift of Good Land (1979)

My purpose here is double. I want, first, to attempt a Biblical argument for ecological and agricultural responsibility. Second, I want to examine some of the practical implications of such an argument. I am prompted to the first of these tasks partly because of its importance in our unresolved conflict about how we should use the world. That those who affirm the divinity of the Creator should come to the rescue of His creature is a logical consistency of great potential force.

The second task is obviously related to the first, but my motive here is somewhat more personal. I wish to deal directly at last with my own long held belief that Christianity, as usually presented by its organizations, is not *earthly* enough—that a valid spiritual life, in this world, must have a practice and a practicality—it must have a material result. (I am well aware that in this belief I am not alone.) What I shall be working toward is some sort of practical understanding of what Arthur O. Lovejoy called the "this-worldly" aspect of Biblical thought. I want to see if there is not at least implicit in the Judeo-

Christian heritage a doctrine such as what the Buddhists call "right livelihood" or "right occupation."

Some of the reluctance to make a forthright Biblical argument against the industrial rape of the natural world seems to come from the suspicion that this rape originates with the Bible, that Christianity cannot cure what, in effect, it has caused. Judging from conversations I have had, the best-known spokesman for this view is Professor Lynn White, Jr., whose essay "The Historical Roots of Our Ecologic Crisis" has been widely published.

Professor White asserts that it is a "Christian axiom that nature has no reason for existence save to serve man." He seems to base his argument on one Biblical passage, Genesis 1:28, in which Adam and Eve are instructed to "subdue" the earth. "Man," says Professor White, "named all the animals, thus establishing his dominance over them." There is no doubt that Adam's superiority over the rest of Creation was represented, if not established, by this act of naming; he *was* given dominance. But that this dominance was meant to be tyrannical, or that "subdue" meant to destroy, is by no means a necessary inference. Indeed, it might be argued that the correct understanding of this "dominance" is given in Genesis 2:15, which says that Adam and Eve were put into the Garden "to dress it and to keep it."

But these early verses of Genesis can give us only limited help. The instruction in Genesis 1:28 was, after all, given to Adam and Eve in the time of their innocence, and it seems certain that the word "subdue" would have had a different intent and sense for them at that time than it could have for them, or for us, after the Fall.

It is tempting to quarrel at length with various statements in Professor White's essay, but he has made that unnecessary by giving us two sentences that define both his problem and my task. He writes, first, that "God planned all of this [the Creation] explicitly for man's benefit and rule: no item in the physical creation had any purpose save to serve man's purposes." And a few sentences later he says: "Christianity . . . insisted that it is God's will that man exploit nature for his *proper* ends" [my emphasis].

It is certainly possible that there might be a critical difference between "man's purposes" and "man's *proper* ends." And one's belief or disbelief in that difference, and one's seriousness about the issue of propriety, will tell a great deal about one's understanding of the Judeo-Christian tradition.

I do not mean to imply that I see no involvement between that tradition and the abuse of nature. I know very well that Christians have not only been often indifferent to such abuse, but have often condoned it and often perpetrated it. That is not the issue. The issue is whether or not the Bible explicitly or implicitly defines a *proper* human use of Creation or the natural world. Proper use, as opposed to improper use, or abuse, is a matter of great complexity, and to find it adequately treated it is necessary to turn to a more complex story than that of Adam and Eve.

The story of the giving of the Promised Land to the Israelites is more serviceable than the story of the giving of the Garden of Eden, because the Promised Land is a divine gift to a *fallen* people. For that reason the giving is more problematical, and the receiving is more conditional and more difficult. In the Bible's long working out of the understanding of this gift, we may find the beginning—and, by implication, the end—of the definition of an ecological discipline.

The effort to make sense of this story involves considerable difficulty because the tribes of Israel, though they see the Promised Land as a gift to them from God, are also obliged to take it by force from its established inhabitants. And so a lot of the "divine sanction" by which they act sounds like the sort of rationalization that invariably accompanies nationalistic aggression and theft. It is impossible to ignore the similarities to the westward movement of the American frontier. The Israelites were following their own doctrine of "manifest destiny," which for them, as for us, disallowed any human standing to their opponents. In Canaan, as in America, the conquerors acted upon the broadest possible definition of idolatry and the narrowest possible definition of justice. They conquered with the same ferocity and with the same genocidal intent.

But for all these similarities, there is a significant difference. Whereas the greed and violence of the American frontier produced an ethic of greed and violence that justified American industrialization, the ferocity of the conquest of Canaan was accompanied from the beginning by the working out of an ethical system antithetical to it—and antithetical, for that matter, to the American conquest with which I have compared it. The difficulty but also the wonder of the story of the Promised Land is that, there, the primordial and still continuing dark story of human rapaciousness begins to be accompanied by a vein of light that, however improbably and uncertainly, still accompanies us. This light originates in the idea of the land as a gift—not a free or a deserved gift, but a gift given upon certain rigorous conditions.

It is a gift because the people who are to possess it did not create it. It is accompanied by careful warnings and demonstrations of the folly of saying that "My power and the might of mine hand hath gotten me this wealth"

(Deuteronomy 8:17). Thus, deeply implicated in the very definition of this gift is a specific warning against *hubris*, which is the great ecological sin, just as it is the great sin of politics. People are not gods. They must not act like gods or assume godly authority. If they do, terrible retributions are in store. In this warning we have the root of the idea of propriety, of *proper* human purposes and ends. We must not use the world as though we created it ourselves.

The Promised Land is not a permanent gift. It is "given," but only for a time, and only for so long as it is properly used. It is stated unequivocally, and repeated again and again, that "the heaven and the heaven of heavens is the Lord's thy God, the earth also, with all that therein is" (Deuteronomy 10:14). What is given is not ownership, but a sort of tenancy, the right of habitation and use: "The land shall not be sold forever: for the land is mine; for ye are strangers and sojourners with me" (Leviticus 25:23).

In token of His landlordship, God required a sabbath for the land, which was to be left fallow every seventh year; and a sabbath of sabbaths every fiftieth year, a "year of jubilee," during which not only would the fields lie fallow, but the land would be returned to its original owners, as if to free it of the taint of trade and the conceit of human ownership. But beyond their agricultural and social intent, these sabbaths ritualize an observance of the limits of "my power and the might of mine hand"—the limits of human control. Looking at their fallowed fields, the people are to be reminded that the land is theirs only by gift; it exists in its own right, and does not begin or end with any human purpose.

The Promised Land, moreover, is "a land which the Lord thy God careth for: the eyes of the Lord thy God are always upon it" (Deuteronomy 11:12). And this care promises a repossession by the true landlord, and a fulfillment not in the power of its human inhabitants: "as truly as I live, all the earth shall be filled with the glory of the Lord" (Numbers 14:21)—a promise recalled by St. Paul in Romans 8:21: "the creature [the Creation] itself also shall be delivered from the bondage of corruption into the glorious liberty of the children of God."

Finally, and most difficult, the good land is not given as a reward. It is made clear that the people chosen for this gift do not deserve it, for they are "a stiff-necked people" who have been wicked and faithless. To such a people such a gift can be given only as a moral predicament: having failed to deserve it beforehand, they must prove worthy of it afterwards; they must use it well, or they will not continue long in it.

How are they to prove worthy?

First of all, they must be faithful, grateful, and humble; they must remember that the land is a gift: "When thou hast eaten and art full, then thou shalt bless the Lord thy God for the good land which he hath given thee" (Deuteronomy 8:10).

Second, they must be neighborly. They must be just, kind to one another, generous to strangers, honest in trading, etc. These are social virtues, but, as they invariably do, they have ecological and agricultural implications. For the land is described as an "inheritance"; the community is understood to exist not just in space, but also in time. One lives in the neighborhood, not just of those who now live "next door," but of the dead who have bequeathed the land to the living, and of the unborn to whom the living will in turn bequeath it. But we can have no direct behavioral connection to those who are not yet alive. The only neighborly thing we can do for them is to preserve their inheritance: we must take care, among other things, of the land, which is never a possession, but an inheritance to the living, as it will be to the unborn.

And so the third thing the possessors of the land must do to be worthy of it is to practice good husbandry. The story of the Promised Land has a good deal to say on this subject, and yet its account is rather fragmentary. We must depend heavily on implication. For sake of brevity, let us consider just two verses (Deuteronomy 22:6–7):

> If a bird's nest chance to be before thee in the way in any tree, or on the ground, whether they be young ones, or eggs, and the dam sitting upon the young, or upon the eggs, thou shalt not take the dam with the young:
>
> But thou shalt in any wise let the dam go, and take the young to thee; that it may be well with thee, and that thou mayest prolong thy days.

This, obviously, is a perfect paradigm of ecological and agricultural discipline, in which the idea of inheritance is necessarily paramount. The inflexible rule is that the source must be preserved. You may take the young, but you must save the breeding stock. You may eat the harvest, but you must save seed, and you must preserve the fertility of the fields.

What we are talking about is an elaborate understanding of charity. It is so elaborate because of the perception, implicit here, explicit in the New Testament, that charity by its nature cannot be selective—that it is, so to speak, out of human control. It cannot be selective because between any two humans, or any two creatures, all Creation exists as a bond. Charity cannot be just human, any more than it can be just Jewish or just Samaritan. Once begun, wherever it begins, it cannot stop until it includes all Creation, for all creatures are parts of a whole upon which each is dependent, and it is a contradiction to love your neighbor and despise the great inheritance on which his life depends. Charity even for one person does not make sense except in terms of an effort to love all Creation in response to the Creator's love for it.

And how is this charity answerable to "man's purposes"? It is not, any more than the Creation itself is. Professor White's contention that the Bible proposes any such thing is, so far as I can see, simply wrong. It is not allowable to love the Creation according to the purposes one has for it, any more than it is allowable to love one's neighbor in order to borrow his tools. The wild ass and the unicorn are said in the Book of Job (39:5–12) to be "free," precisely in the sense that they are not subject or serviceable to human purposes. The same point—though it is not the main point of that passage—is made in the Sermon on the Mount in reference to "the fowls of the air" and "the lilies of the field." Faced with this problem in Book VIII of *Paradise Lost,* Milton scrupulously observes the same reticence. Adam asks about "celestial Motions," and Raphael refuses to explain, making the ultimate mysteriousness of Creation a test of intellectual propriety and humility:

> . . . for the Heav'n's wide Circuit, let it speak
> The Maker's high magnificence, who built
> So spacious, and his Line stretcht out so far;
> That Man may know he dwells not in his own;
> An Edifice too large for him to fill,
> Lodg'd in a small partition, and the rest
> Ordain'd for uses to his Lord best known.
>
> (lines 100–106)

The Creator's love for the Creation is mysterious precisely because it does not conform to human purposes. The wild ass and the wild lilies are loved by God for their own sake and yet they are part of a pattern that we must love because it includes us. This is a pattern that humans can understand well enough to respect and preserve, though they cannot "control" it or hope to understand it completely. The mysterious and the practical, the Heavenly and the earthly, are thus joined. Charity is a theological virtue and is prompted, no doubt, by a theological emotion, but it is also a practical virtue because it must be practiced. The requirements of this complex charity cannot be fulfilled by smiling in abstract beneficence on our neighbors and on the scenery. It must come to acts, which must come from skills. Real charity calls for the study of agriculture, soil husbandry, engineering, architecture, mining, manufacturing, transportation, the making of monuments and pictures, songs and stories. It calls not just for skills but for the study and criticism of skills, because in all of them a choice must be made: they can be used either charitably or uncharitably.

How can you love your neighbor if you don't know how to build or mend a fence, how to keep your filth out of his water supply and your poison out of his air; or if you do not produce anything and so have nothing to offer, or do not take care of yourself and so become a burden? How can you be a neighbor without *applying* principle—without bringing virtue to a practical issue? How will you practice virtue without skill?

The ability to be good is not the ability to do nothing. It is not negative or passive. It is the ability to do something well—to do good work for good reasons. In order to be good you have to know how—and this knowing is vast, complex, humble and humbling; it is of the mind and of the hands, of neither alone.

The divine mandate to use the world justly and charitably, then, defines every person's moral predicament as that of a steward. But this predicament is hopeless and meaningless unless it produces an appropriate discipline: stewardship. And stewardship is hopeless and meaningless unless it involves long-term courage, perseverance, devotion, and skill. This skill is not to be confused with any accomplishment or grace of spirit or of intellect. It has to do with everyday proprieties in the practical use and care of created things—with "right livelihood."

If "the earth is the Lord's" and we are His stewards, then obviously some livelihoods are "right" and some are not. Is there, for instance, any such thing as a Christian strip mine? A Christian atomic bomb? A Christian nuclear power plant or radioactive waste dump? What might be the design of a Christian transportation or sewer system? Does not Christianity imply limitations on the scale of technology, architecture, and land holding? Is it Christian to profit or otherwise benefit from violence? Is there not, in Christian ethics, an implied requirement of practical separation from a destructive or wasteful economy? Do not Christian values require the enactment of a distinction between an organization and a community?

It is impossible to understand, much less to answer, such questions except in reference to issues of practical skill, because they all have to do with distinctions between kinds of action. These questions, moreover, are intransigently personal, for they ask, ultimately, how each livelihood and each life will be taken from the world, and what each will cost in terms of the livelihoods and lives of others. Organizations and even communities cannot hope to answer such questions until individuals have begun to answer them.

But here we must acknowledge one inadequacy of Judeo-Christian tradition. At least in its most prominent and best-known examples, this tradition does not provide us with a precise enough understanding of the commonplace issues of livelihood. There are two reasons for this.

One is the "otherworldly philosophy" that, according to Lovejoy, "has, in one form or another, been the dominant official philosophy of the larger part of civilized mankind through most of its history. . . . The greater number of the subtler speculative minds and of the great religious teachers have . . . been engaged in weaning man's thought or his affections, or both, from . . . Nature." The connection here is plain.

The second reason is that the Judeo-Christian tradition as we have it in its art and literature, including the Bible, is so strongly heroic. The poets and storytellers in this tradition have tended to be interested in the extraordinary actions of "great men"—actions unique in grandeur, such as may occur only once in the history of the world. These extraordinary actions do indeed bear a universal significance, but they cannot very well serve as examples of ordinary behavior. Ordinary behavior belongs to a different dramatic mode, a different understanding of action, even a different understanding of virtue. The drama of heroism raises above all the issue of physical and moral courage: Does the hero have, in extreme circumstances, the courage to obey—to perform the task, the sacrifice, the resistance, the pilgrimage that he is called on to perform? The drama of ordinary or daily behavior also raises the issue of courage, but it raises at the same time the issue of skill; and, because ordinary behavior lasts so much longer than heroic action, it raises in a more complex and difficult way the issue of perseverance. It may, in some ways, be easier to be Samson than to be a good husband or wife day after day for fifty years.

These heroic works are meant to be (among other things) instructive and inspiring to ordinary people in ordinary life, and they are, grandly and deeply so. But there are two issues that they are prohibited by their nature from raising: the issue of lifelong devotion and perseverance in unheroic tasks, and the issue of good workmanship or "right livelihood."

It can be argued, I believe, that until fairly recently there was simply no need for attention to such issues, for there existed yeoman or peasant or artisan classes: these were the people who did the work of feeding and clothing and housing, and who were responsible for the necessary skills, disciplines, and restraints. As long as those earth-keeping classes and their traditions were strong, there was at least the hope that the world would be well used. But probably the most revolutionary accomplishment of the industrial revolution was to destroy the traditional livelihoods and so break down the cultural lineage of those classes.

The industrial revolution has held in contempt not only the "obsolete skills" of those classes, but the concern for quality, for responsible workmanship and good work, that supported their skills. For the principle of good work it substituted a secularized version of the heroic tradition: the ambition to be a "pioneer" of science or technology, to make a "breakthrough" that will "save the world" from some "crisis" (which now is usually the result of some previous "breakthrough").

The best example we have of this kind of hero, I am afraid, is the fallen Satan of *Paradise Lost*—Milton undoubtedly having observed in his time the prototypes of industrial heroism. This is a hero who instigates and influences the actions of others, but does not act himself. His heroism is of the mind only—escaped as far as possible, not only from divine rule, from its place in the order of creation or the Chain of Being, but also from the influence of material creation:

> A mind not to be chang'd by Place or Time.
> The mind is its own place, and in itself
> Can make a Heav'n of Hell, a Hell of Heav'n.
>
> (Book I, lines 253–255)

This would-be heroism is guilty of two evils that are prerequisite to its very identity: *hubris* and abstraction. The industrial hero supposes that "mine own *mind* hath saved me"—and moreover that it may save the world. Implicit in this is the assumption that one's mind is one's own, and that it may choose its own place in the order of things; one usurps divine authority, and thus, in classic style, becomes the author of results that one can neither foresee nor control.

And because this mind is understood only as a cause, its primary works are necessarily abstract. We should remind ourselves that materialism in the sense of the love of material things is not in itself an evil. As C. S. Lewis pointed out, God too loves material things; He invented them. The Devil's work is abstraction—not the love of material things, but the love of their quantities—which, of course, is why "David's heart smote him after that he had numbered the people" (II Samuel 24:10). It is not the lover of material things but the abstractionist who defends long-term damage for short-term gain, or who calculates the "acceptability" of industrial damage to ecological or human health, or who counts dead bodies on the battlefield. The true lover of material things does not think in this way, but is answerable instead to the paradox of the parable of the lost sheep: that each is more precious than all.

But perhaps we cannot understand this secular heroic mind until we understand its opposite: the mind obedient and in place. And for that we can look again at Raphael's warning in Book VIII of *Paradise Lost:*

> . . . apt the Mind or Fancy is to rove
> Uncheckt, and of her roving is no end;
> Till warn'd, or by experience taught, she learn
> That not to know at large of things remote
> From use, obscure and subtle, but to know
> That which before us lies in daily life,
> Is the prime Wisdom; what is more, is fume,
> Or emptiness, or fond impertinence,
> And renders us in things that most concern
> Unpractic'd, unprepar'd, and still to seek.
> Therefore from this high pitch let us descend
> A lower flight, and speak of things at hand
> Useful . . .
>
> (lines 188–200)

In its immediate sense this is a warning against thought that is theoretical or speculative (and therefore abstract), but in its broader sense it is a warning against disobedience—the eating of the forbidden fruit, an act of *hubris*, which Satan justifies by a compellingly reasonable theory and which Eve undertakes as a speculation.

A typical example of the conduct of industrial heroism is to be found in the present rush of experts to "solve the problem of world hunger"—which is rarely defined except as a "world problem" known, in industrial heroic jargon, as "the world food problematique." As is characteristic of industrial heroism, the professed intention here is entirely salutary: nobody should starve. The trouble is that "world hunger" is not a problem that can be solved by a "world solution." Except in a very limited sense, it is not an industrial problem, and industrial attempts to solve it—such as the "Green Revolution" and "Food for Peace"—have often had grotesque and destructive results. "The problem of world hunger" cannot be solved until it is understood and dealt with by local people as a multitude of local problems of ecology, agriculture, and culture.

The most necessary thing in agriculture, for instance, is not to invent new technologies or methods, not to achieve "breakthroughs," but to determine what tools and methods are appropriate to specific people, places, and needs, and to apply them correctly. Application (which the heroic approach ignores) is the crux, because no two farms or farmers are alike; no two fields are alike. Just the changing shape or topography of the land makes for differences of the most formidable kind. Abstractions never cross these boundaries without either ceasing to be abstractions or doing damage. And prefabricated industrial methods and technologies *are* abstractions. The bigger and more expensive, the more heroic, they are, the harder they are to apply considerately and conservingly.

Application is the most important work, but also the most modest, complex, difficult, and long—and so it goes against the grain of industrial heroism. It destroys forever the notions that the world can be thought of (by humans) as a whole and that humans can "save" it as a whole—notions we can well do without, for they prevent us from understanding our problems and from growing up.

To use knowledge and tools in a particular place with good long-term results is not heroic. It is not a grand action visible for a long distance or a long time. It is a small action, but more complex and difficult, more skillful and responsible, more whole and enduring, than most grand actions. It comes of a willingness to devote oneself to work that perhaps only the eye of Heaven will see in its full intricacy and excellence. Perhaps the real work, like real prayer and real charity, must be done in secret.

The great study of stewardship, then, is "to know / That which before us lies in daily life" and to be practiced and prepared "in things that most concern." The angel is talking about good work, which is to talk about skill. In the loss of skill we lose stewardship; in losing stewardship we lose fellowship; we become outcasts from the great neighborhood of Creation. It is possible—as our experience in *this* good land shows—to exile ourselves from Creation, and to ally ourselves with the principle of destruction—which is, ultimately, the principle of nonentity. It is to be willing *in general* for beings to not-be. And once we have allied ourselves with that principle, we are foolish to think that we can control the results. The "regulation" of abominations is a modern governmental exercise that never succeeds. If we are willing to pollute the air—to harm the elegant creature known as the atmosphere—by that token we are willing to harm all creatures that breathe, ourselves and our children among them. There is no begging off or "trading off." You cannot affirm the power plant and condemn the smokestack, or affirm the smoke and condemn the cough.

That is not to suggest that we can live harmlessly, or strictly at our own expense; we depend upon other creatures and survive by their deaths. To live, we must daily break the body and shed the blood of Creation. When we do this knowingly, lovingly, skillfully, reverently, it is a sacrament. When we do it ignorantly, greedily, clumsily, destructively, it is a desecration. In such desecration we condemn ourselves to spiritual and moral loneliness, and others to want.

Christianity and the Survival of Creation*

I

I want to begin with a problem: namely, that the culpability of Christianity in the destruction of the natural world and the uselessness of Christianity in any effort to correct that destruction are now established clichés of the conservation movement. This is a problem for two reasons.

First, the indictment of Christianity by the anti-Christian conservationists is, in many respects, just. For instance, the complicity of Christian priests, preachers, and missionaries in the cultural destruction and the economic exploitation of the primary peoples of the Western Hemisphere, as of traditional cultures around the world, is notorious. Throughout the five hundred years since Columbus's first landfall in the Bahamas, the evangelist has walked beside the conqueror and the merchant, too often blandly assuming that their causes were the same. Christian organizations, to this day, remain largely

*This essay was delivered as a lecture at the Southern Baptist Theological Seminary in Louisville, Kentucky.

indifferent to the rape and plunder of the world and of its traditional cultures. It is hardly too much to say that most Christian organizations are as happily indifferent to the ecological, cultural, and religious implications of industrial economics as are most industrial organizations. The certified Christian seems just as likely as anyone else to join the military-industrial conspiracy to murder Creation.

The conservationist indictment of Christianity is a problem, second, because, however just it may be, it does not come from an adequate understanding of the Bible and the cultural traditions that descend from the Bible. The anti-Christian conservationists characteristically deal with the Bible by waving it off. And this dismissal conceals, as such dismissals are apt to do, an ignorance that invalidates it. The Bible is an inspired book written by human hands; as such, it is certainly subject to criticism. But the anti-Christian environmentalists have not mastered the first rule of the criticism of books: you have to read them before you criticize them. Our predicament now, I believe, requires us to learn to read and understand the Bible in the light of the present fact of Creation. This would seem to be a requirement both for Christians and for everyone concerned, but it entails a long work of true criticism — that is, of careful and judicious study, not dismissal. It entails, furthermore, the making of very precise distinctions between biblical instruction and the behavior of those peoples supposed to have been biblically instructed.

I cannot pretend, obviously, to have made so meticulous a study; even if I were capable of it, I would not live long enough to do it. But I have attempted to read the Bible with these issues in mind, and I see some virtually catastrophic discrepancies between biblical instruction and Christian behavior. I don't mean disreputable Christian behavior, either. The discrepancies I see are between biblical instruction and allegedly respectable Christian behavior.

If because of these discrepancies Christianity were dismissible, there would, of course, be no problem. We could simply dismiss it, along with the twenty centuries of unsatisfactory history attached to it, and start setting things to rights. The problem emerges only when we ask, Where then would we turn for instruction? We might, let us suppose, turn to another religion — a recourse that is sometimes suggested by the anti-Christian conservationists. Buddhism, for example, is certainly a religion that could guide us toward a right respect for the natural world, our fellow humans, and our fellow creatures. I owe a considerable debt myself to Buddhism and Buddhists. But there are an enormous number of people — and I am one of them — whose native religion, for better or worse, is Christianity. We were born to it; we began to learn about it before we became conscious; it is, whatever we think of it, an intimate belonging of our being; it informs our consciousness, our language, and our dreams. We can turn away from it or against it, but that will only bind us tightly to a reduced version of it. A better possibility is that this, our native religion, should survive and renew itself so that it may become as largely and truly instructive as we need it to be. On such a survival and renewal of the Christian religion may depend the survival of the Creation that is its subject.

II

If we read the Bible, keeping in mind the desirability of those two survivals—of Christianity and the Creation—we are apt to discover several things about which modern Christian organizations have kept remarkably quiet or to which they have paid little attention.

We will discover that we humans do not own the world or any part of it: "The earth is the Lord's, and the fulness thereof: the world and they that dwell therein." There is in our human law, undeniably, the concept and right of "land ownership." But this, I think, is merely an expedient to safeguard the mutual belonging of people and places without which there can be no lasting and conserving human communities. This right of human ownership is limited by mortality and by natural constraints on human attention and responsibility; it quickly becomes abusive when used to justify large accumulations of "real estate," and perhaps for that reason such large accumulations are forbidden in the twenty-fifth chapter of Leviticus. In biblical terms, the "landowner" is the guest and steward of God: "The land is mine; for ye are strangers and sojourners with me."

We will discover that God made not only the parts of Creation that we humans understand and approve but all of it: "All things were made by him; and without him was not anything made that was made." And so we must credit God with the making of biting and stinging insects, poisonous serpents, weeds, poisonous weeds, dangerous beasts, and disease-causing microorganisms. That we may disapprove of these things does not mean that God is in error or that He ceded some of the work of Creation to Satan; it means that we are deficient in wholeness, harmony, and understanding—that is, we are "fallen."

We will discover that God found the world, as He made it, to be good, that He made it for His pleasure, and that He continues to love it and to find it worthy, despite its reduction and corruption by us. People who quote John 3:16 as an easy formula for getting to Heaven neglect to see the great difficulty implied in the statement that the advent of Christ was made possible by God's love for the world — not God's love for Heaven or for the world as it might be but for the world as it was and is. Belief in Christ is thus dependent on prior belief in the inherent goodness — the lovability — of the world.

We will discover that the Creation is not in any sense independent of the Creator, the result of a primal creative act long over and done with, but is the continuous, constant participation of all creatures in the being of God. Elihu said to Job that if God "gather unto himself his spirit and his breath; all flesh shall perish together." And Psalm 104 says, "Thou sendest forth thy spirit, they are created." Creation is thus God's presence in creatures. The Greek Orthodox theologian Philip Sherrard has written that "Creation is nothing less than the manifestation of God's hidden Being." This means that we and all other creatures live by a sanctity that is inexpressibly intimate, for to every creature, the gift of life is a portion of the breath and spirit of God. As the poet George Herbert put it:

> Thou art in small things great, not small in any . . .
> For thou art infinite in one and all.

We will discover that for these reasons our destruction of nature is not just bad stewardship, or stupid economics, or a betrayal of family responsibility; it is the most horrid blasphemy. It is flinging God's gifts into His face, as if they were of no worth beyond that assigned to them by our destruction of them. To Dante, "despising Nature and her goodness" was a violence against God. We have no entitlement from the Bible to exterminate or permanently destroy or hold in contempt anything on the earth or in the heavens above it or in the waters beneath it. We have the right to use the gifts of nature but not to ruin or waste them. We have the right to use what we need but no more, which is why the Bible forbids usury and great accumulations of property. The usurer, Dante said, "condemns Nature . . . for he puts his hope elsewhere."

William Blake was biblically correct, then, when he said that "everything that lives is holy." And Blake's great commentator Kathleen Raine was correct both biblically and historically when she said that "the sense of the holiness of life is the human norm."

The Bible leaves no doubt at all about the sanctity of the act of world-making, or of the world that was made, or of creaturely or bodily life in this world. We are holy creatures living among other holy creatures in a world that is holy. Some people know this, and some do not. Nobody, of course, knows it all the time. But what keeps it from being far better known than it is? Why is it apparently unknown to millions of professed students of the Bible? How can modern Christianity have so solemnly folded its hands while so much of the work of God was and is being destroyed?

III

Obviously, "the sense of the holiness of life" is not compatible with an exploitive economy. You cannot know that life is holy if you are content to live from economic practices that daily destroy life and diminish its possibility. And many if not most Christian organizations now appear to be perfectly at peace with the military-industrial economy and its "scientific" destruction of life. Surely, if we are to remain free and if we are to remain true to our religious inheritance, we must maintain a separation between church and state. But if we are to maintain any sense or coherence or meaning in our lives, we cannot tolerate the present utter disconnection between religion and economy. By "economy" I do not mean "economics," which is the study of money-making, but rather the ways of human housekeeping, the ways by which the human household is situated and maintained within the household of nature. To be uninterested in economy is to be uninterested in the practice of religion; it is to be uninterested in culture and in character. Probably the most urgent question now faced by people who would adhere to the Bible is this: What sort of economy would be responsible to the holiness of life? What, for Christians, would be the economy, the practices and the restraints, of "right livelihood"? I do not believe that organized Christianity now has any idea. I think its idea of a Christian economy is no more or less than the industrial economy—which is an economy firmly founded on the seven deadly sins and the breaking of all ten of the Ten Commandments. Obviously, if Christianity is going to survive as more than a respecter and comforter of profitable iniquities, then Christians, regardless of their organizations, are going to have to interest themselves in economy—which is to say, in nature and in work. They are going to have to give workable answers to those who say we cannot live without this economy that is destroying us and our world, who see the murder of Creation as the only way of life.

The holiness of life is obscured to modern Christians also by the idea that the only holy place is the built church. This idea may be more taken for granted than taught; nevertheless, Christians are encouraged from childhood to think of the church building as "God's house," and most of them could think of their houses or farms or shops or factories as holy places only with great effort and embarrassment. It is understandably difficult for modern Americans to think of their dwellings and workplaces as holy, because most of these are, in fact, places of desecration, deeply involved in the ruin of Creation.

The idea of the exclusive holiness of church buildings is, of course, wildly incompatible with the idea, which the churches also teach, that God is present in all places to hear prayers. It is incompatible with Scripture. The idea that a human artifact could contain or confine God was explicitly repudiated by Solomon in his prayer at the dedication of the Temple: "Behold, the heaven and the heaven of heavens cannot contain thee: how much less this house that I have builded?" And these words of Solomon were remembered a thousand years later by Saint Paul, preaching at Athens:

> God that made the world and all things therein, seeing that he is lord of heaven and earth, dwelleth not in temples made with hands . . .
>
> For in him we live, and move, and have our being; as certain also of your own poets have said.

Idolatry always reduces to the worship of something "made with hands," something confined within the terms of human work and human comprehension. Thus, Solomon and Saint Paul both insisted on the largeness and the at-largeness of God, setting Him free, so to speak, from *ideas* about Him. He is not to be fenced in, under human control, like some domestic creature; He is the wildest being in existence. The presence of His spirit in us is our wildness, our oneness with the wilderness of Creation. That is why subduing the things of nature to human purposes is so dangerous and why it so often results in evil, in separation and desecration. It is why the poets of our tradition so often have given nature the role not only of mother or grandmother but of the highest earthly teacher and judge, a figure of mystery and great power. Jesus' own specifications for his church have nothing at all to do with masonry and carpentry but only with people; his church is "where two or three are gathered together in my name."

The Bible gives exhaustive (and sometimes exhausting) attention to the organization of religion: the building and rebuilding of the Temple; its furnishings; the orders, duties, and paraphernalia of the priesthood; the orders of rituals and ceremonies. But that does not disguise the fact that the most significant religious events recounted in the Bible do not occur in "temples made with hands." The most important religion in that book is unorganized and is sometimes profoundly disruptive of organization. From Abraham to Jesus, the most important people are not priests but shepherds, soldiers, property owners, workers, housewives, queens and kings, manservants and maidservants, fishermen, prisoners, whores, even bureaucrats. The great visionary encounters did not take place in temples but in sheep pastures, in the desert, in the wilderness, on mountains, on the shores of rivers and the sea, in the middle of the sea, in prisons. And however strenuously the divine voice prescribed rites and observances, it just as strenuously repudiated them when they were taken to *be* religion:

> Your new moons and your appointed feasts my soul hateth: they are a trouble unto me; I am weary to bear them.
>
> And when you spread forth your hands, I will hide mine eyes from you: yea, when you make many prayers, I will not hear: your hands are full of blood.
>
> Wash you, make you clean; put away the evil of your doings from before mine eyes; cease to do evil;
>
> Learn to do well; seek judgment, relieve the oppressed, judge the fatherless, plead for the widow.

Religion, according to this view, is less to be celebrated in rituals than practiced in the world.

I don't think it is enough appreciated how much an outdoor book the Bible is. It is a "hypaethral book," such as Thoreau talked about—a book open to the sky. It is best read and understood outdoors, and the farther outdoors the better. Or that has been my experience of it. Passages that within walls seem improbable or incredible, outdoors seem merely natural. This is because outdoors we are confronted everywhere with wonders; we see that the miraculous is not extraordinary but the common mode of existence. It is our daily bread. Whoever really has considered the lilies of the field or the birds of the air and pondered the improbability of their existence in this warm world within the cold and empty stellar distances will hardly balk at the turning of water into wine—which was, after all, a very small miracle. We forget the greater and still continuing miracle by which water (with soil and sunlight) is turned into grapes.

It is clearly impossible to assign holiness exclusively to the built church without denying holiness to the rest of Creation, which is then said to be "secular." The world, which God looked at and found entirely good, we find none too good to pollute entirely and destroy piecemeal. The church, then, becomes a kind of preserve of "holiness," from which certified lovers of God assault and plunder the "secular" earth.

Not only does this repudiate God's approval of His work; it refuses also to honor the Bible's explicit instruction to regard the works of the Creation as God's revelation of Himself. The assignation of holiness exclusively to the built church is therefore logically accompanied by the assignation of revelation exclusively to the Bible. But Psalm 19 begins, "The heavens declare the glory of God; and the firmament sheweth his handiwork." The word of God has been revealed in facts from the moment of the third verse of the first chapter of Genesis: "Let there be light: and there was light." And Saint Paul states the rule: "The invisible things of him from the creation of the world are clearly seen, being understood by the things that are made." Yet from this free, generous, and sensible view of things, we come to the idolatry of the book: the idea that nothing is true that cannot be (and has not been already) written. The misuse of the Bible thus logically accompanies the abuse of nature: if you are going to destroy creatures without respect, you will want to reduce them to "materiality"; you will want to deny that there is spirit or truth in them, just as you will want to believe that the only holy creatures, the only creatures with souls, are humans—or even only Christian humans.

By denying spirit and truth to the nonhuman Creation, modern proponents of religion have legitimized a form of blasphemy without which the nature- and culture-destroying machinery of the industrial economy could not have been built—that is, they have legitimized bad work. Good human work honors God's work. Good work uses no thing without respect, both for what it is in itself and for its origin. It uses neither tool nor material that it does not respect and that it does not love. It honors nature as a great mystery and power, as an indispensable teacher, and as the inescapable judge of all work of human hands. It does not dissociate life and work, or pleasure and work, or love and work, or usefulness and beauty. To work without pleasure or affection, to make a product that is not both useful and beautiful, is to dishonor God, nature, the thing that is made, and whomever it is made for. This is blasphemy: to make shoddy work of the work of God. But such blasphemy is not possible when the entire Creation is understood as holy and when the works of God are understood as embodying and thus revealing His spirit.

In the Bible we find none of the industrialist's contempt or hatred for nature. We find, instead, a poetry of awe and reverence and profound cherishing, as in these verses from Moses' valedictory blessing of the twelve tribes:

> And of Joseph he said, Blessed of the Lord be his land, for the precious things of heaven, for the dew, and for the deep that croucheth beneath,
>
> And for the precious fruits brought forth by the sun, and for the precious things put forth by the moon,
>
> And for the chief things of the ancient mountains, and for the precious things of the lasting hills,
>
> And for the precious things of the earth and fullness thereof, and for the good will of him that dwelt in the bush.

IV

I have been talking, of course, about a dualism that manifests itself in several ways: as a cleavage, a radical discontinuity, between Creator and creature, spirit and matter, religion and nature, religion and economy, worship and work, and so on. This dualism, I think, is the most destructive disease that afflicts us. In its best-known, its most dangerous, and perhaps its fundamental version, it is the dualism of body and soul. This is an issue as difficult as it is important, and so to deal with it we should start at the beginning.

The crucial test is probably Genesis 2:7, which gives the process by which Adam was created: "The Lord God formed man of the dust of the ground, and breathed into his nostrils the breath of life: and man became a living soul." My mind, like most people's, has been deeply influenced by dualism, and I can see how dualistic minds deal with this verse. They conclude that the formula for man-making is man = body + soul. But that conclusion cannot be derived, except by violence, from Genesis 2:7, which is not dualistic. The formula given in Genesis 2:7 is not man = body + soul; the formula there is soul = dust + breath. According to this verse, God did not make a body and put a soul into it, like a letter into an envelope. He formed man of dust; then, by breathing His breath into it, He made the dust live. The dust, formed as man and made to live, did not *embody* a soul; it *became* a soul. "Soul" here refers to the whole creature. Humanity is thus presented to us, in Adam, not as a creature of two discrete parts temporarily glued together but as a single mystery.

We can see how easy it is to fall into the dualism of body and soul when talking about the inescapable worldly dualities of good and evil or time and eternity. And we can see how easy it is, when Jesus asks, "For what is a man profited, if he shall gain the whole world, and lose his own soul?" to assume that he is condemning the world and appreciating the disembodied soul. But if we give to "soul" here the sense that it has in Genesis 2:7, we see that he is doing no such thing. He is warning that in pursuit of so-called material possessions, we can lose our understanding of ourselves as "living souls"—that is, as creatures of God, members of the holy community of Creation. We can lose the possibility of the atonement of that membership. For we are free, if we choose, to make a duality of our one living soul by disowning the breath of God that is our fundamental bond with one another and with other creatures.

But we can make the same duality by disowning the dust. The breath of God is only one of the divine gifts that make us living souls; the other is the dust. Most of our modern troubles come from our misunderstanding and misvaluation of this dust. Forgetting that the dust, too, is a creature of the Creator, made by the sending forth of His spirit, we have presumed to decide that the dust is "low." We have presumed to say that we are made of two parts: a body and a soul, the body being "low" because made of dust, and the soul "high." By thus valuing these two supposed-to-be parts, we inevitably throw them into competition with each other, like two corporations. The "spiritual" view, of course, has been that the body, in Yeats's phrase, must be "bruised to pleasure soul." And the "secular" version of the same dualism has been that the body, along with the rest of the "material" world, must give way before the advance of the human mind. The dominant religious view, for a long time, has been that the body is a kind of scrip issued by the Great Company Store in the Sky, which can be cashed in to redeem the soul but is otherwise worthless. And the predictable result has been a human creature able to appreciate or tolerate only the "spiritual" (or mental) part of Creation and full of semiconscious hatred of the "physical" or "natural" part, which it is ready and willing to destroy for "salvation," for profit, for "victory," or for fun. This madness constitutes the norm of modern humanity and of modern Christianity.

But to despise the body or mistreat it for the sake of the "soul" is not just to burn one's house for the insurance, nor is it just self-hatred of the most deep and dangerous sort. It is yet another blasphemy. It is to make nothing—and worse than nothing—of the great Something in which we live and move and have our being.

When we hate and abuse the body and its earthly life and joy for Heaven's sake, what do we expect? That out of this life that we have presumed to despise and this world that we have presumed to destroy, we would somehow salvage a soul capable of eternal bliss? And what do we expect when with equal and opposite ingratitude, we try to make of the finite body an infinite reservoir of dispirited and meaningless pleasures?

Times may come, of course, when the life of the body must be denied or sacrificed, times when the whole world must literally be lost for the sake of one's life as a "living soul." But such sacrifice, by people who truly respect and revere the life of the earth and its Creator, does not denounce or degrade the body but rather exalts it and acknowledges its holiness. Such sacrifice is a refusal to allow the body to serve what is unworthy of it.

V

If we credit the Bible's description of the relationship between Creator and Creation, then we cannot deny the spiritual importance of our economic life. Then we must see how religious issues lead to issues of economy and how issues of economy lead to issues of art. By "art" I mean all the ways by which humans make the things they need. If we understand that no artist—no maker—can work except by reworking the works of Creation, then we see that by our work we reveal what we think of the works of God. How we take our lives from this world, how we work, what work we do, how well we use the materials we use, and what we do with them after we have used them—all these are questions of the highest and gravest religious significance. In answering them, we practice, or do not practice, our religion.

The significance—and ultimately the quality—of the work we do is determined by our understanding of the story in which we are taking part.

If we think of ourselves as merely biological creatures, whose story is determined by genetics or environment or history or economics or technology, then, however pleasant or painful the part we play, it cannot matter much. Its significance is that of mere self-concern. "It is a tale / Told by an idiot, full of sound and fury, / Signifying nothing," as Macbeth says when he has "supp'd full with horrors" and is "aweary of the sun."

If we think of ourselves as lofty souls trapped temporarily in lowly bodies in a dispirited, desperate, unlovable world that we must despise for Heaven's sake, then what have we done for this question of significance? If we divide reality into two parts, spiritual and material, and hold (as the Bible does *not* hold) that only the spiritual is good or desirable, then our relation to the material Creation becomes arbitrary, having only the quantitative or mercenary value that we have, in fact and for this reason, assigned to it. Thus, we become the judges and inevitably the destroyers of a world we did not make and that we are bidden to understand as a divine gift. It is impossible to see how good work might be accomplished by people who think that our life in this world either signifies nothing or has only a negative significance.

If, on the other hand, we believe that we are living souls, God's dust and God's breath, acting our parts among other creatures all made of the same dust and breath as ourselves; and if we understand that we are free, within the obvious limits of mortal human life, to do evil or good to ourselves and to the other creatures — then all our acts have a supreme significance. If it is true that we are living souls and morally free, then all of us are artists. All of us are makers, within mortal terms and limits, of our lives, of one another's lives, of things we need and use.

This, Ananda Coomaraswamy wrote, is "the normal view," which "assumes . . . not that the artist is a special kind of man, but that every man who is not a mere idler or parasite is necessarily some special kind of artist." But since even mere idlers and parasites may be said to work inescapably, by proxy or influence, it might be better to say that everybody is an artist — either good or bad, responsible or irresponsible. Any life, by working or not working, by working well or poorly, inescapably changes other lives and so changes the world. This is why our division of the "fine arts" from "craftsmanship," and "craftsmanship" from "labor," is so arbitrary, meaningless, and destructive. As Walter Shewring rightly said, both "the plowman and the potter have a cosmic function." And bad art in any trade dishonors and damages Creation.

If we think of ourselves as living souls, immortal creatures, living in the midst of a Creation that is mostly mysterious, and if we see that everything we make or do cannot help but have an everlasting significance for ourselves, for others, and for the world, then we see why some religious teachers have understood work as a form of prayer. We see why the old poets invoked the muse. And we know why George Herbert prayed, in his poem "Mattens":

> Teach me thy love to know;
> That this new light, which now I see,
> May both the work and workman show.

Work connects us both to Creation and to eternity. This is the reason also for Mother Ann Lee's famous instruction: "Do all your work as though you had a thousand years to live on earth, and as you would if you knew you must die tomorrow."

Explaining "the perfection, order, and illumination" of the artistry of Shaker furniture makers, Coomaraswamy wrote, "All tradition has seen in the

Master Craftsman of the Universe the exemplar of the human artist or 'maker by art,' and we are told to be 'perfect, *even as* your Father in heaven is perfect.'" Searching out the lesson, for us, of the Shakers' humble, impersonal, perfect artistry, which refused the modern divorce of utility and beauty, he wrote, "Unfortunately, we do not desire to be such as the Shaker was; we do not propose to 'work as though we had a thousand years to live, and as though we were to die tomorrow.' Just as we desire peace but not the things that make for peace, so we desire art but not the things that make for art. . . . we have the art that we deserve. If the sight of it puts us to shame, it is with ourselves that the reformation must begin."

Any genuine effort to "re-form" our arts, our ways of making, must take thought of "the things that make for art." We must see that no art begins in itself; it begins in other arts, in attitudes and ideas antecedent to any art, in nature, and in inspiration. If we look at the great artistic traditions, as it is necessary to do, we will see that they have never been divorced either from religion or from economy. The possibility of an entirely secular art and of works of art that are spiritless or ugly or useless is not a possibility that has been among us for very long. Traditionally, the arts have been ways of making that have placed a just value on their materials or subjects, on the uses and the users of the things made by art, and on the artists themselves. They have, that is, been ways of giving honor to the works of God. The great artistic traditions have had nothing to do with what we call "self-expression." They have not been destructive of privacy or exploitive of private life. Though they have certainly originated things and employed genius, they have no affinity with the modern cults of originality and genius. Coomaraswamy, a good guide as always, makes an indispensable distinction between genius in the modern sense and craftsmanship: "Genius inhabits a world of its own. The master craftsman lives in a world inhabited by other men; he has neighbors." The arts, traditionally, belong to the neighborhood. They are the means by which the neighborhood lives, works, remembers, worships, and enjoys itself.

But most important of all, now, is to see that the artistic traditions understood every art primarily as a skill or craft and ultimately as a service to fellow creatures and to God. An artist's first duty, according to this view, is technical. It is assumed that one will have talents, materials, subjects—perhaps even genius or inspiration or vision. But these are traditionally understood not as personal properties with which one may do as one chooses but as gifts of God or nature that must be honored in use. One does not dare to use these things without the skill to use them well. As Dante said of his own art, "far worse than in vain does he leave the shore . . . who fishes for the truth and has not the art." To use gifts less than well is to dishonor them and their Giver. There is no material or subject in Creation that, in using, we are excused from using well; there is no work in which we are excused from being able and responsible artists.

VI

In denying the holiness of the body and of the so-called physical reality of the world—and in denying support to the good economy, the good work, by which alone the Creation can receive due honor—modern Christianity generally has cut itself off from both nature and culture. It has no serious or competent interest in biology or ecology. And it is equally uninterested in the arts by which humankind connects itself to nature. It manifests no awareness of the specifically Christian cultural lineages that connect us to our past. There is, for example, a splendid heritage of Christian poetry in English that most church members live and die without reading or hearing or hearing about. Most sermons are preached without any awareness at all that the making of sermons is an art that has at times been magnificent. Most modern churches look like they were built by robots without reference to the heritage of church architecture or respect for the place; they embody no awareness that work can be worship. Most religious music now attests to the general assumption that religion is no more than a vaguely pious (and vaguely romantic) emotion.

Modern Christianity, then, has become as specialized in its organizations as other modern organizations, wholly concentrated on the industrial shibboleths of "growth," counting its success in numbers, and on the very strange enterprise of "saving" the individual, isolated, and disembodied soul. Having witnessed and abetted the dismemberment of the households, both human and natural, by which we have our being as creatures of God, as living souls, and having made light of the great feast and festival of Creation to which we were bidden as living souls, the modern church presumes to be able to save the soul as an eternal piece of private property. It presumes moreover to save the souls of people in other countries and religious traditions, who are often saner and more religious than we are. And always the emphasis is on the individual soul. Some Christian spokespeople give the impression that the highest Christian bliss would be to get to Heaven and find that you are the only one there—that you were right and all the others wrong. Whatever its twentieth-century dress, modern Christianity as I know it is still at bottom the religion of Miss Watson, intent on a dull and superstitious rigmarole by which supposedly we can avoid going to "the bad place" and instead go to "the good place." One can hardly help sympathizing with Huck Finn when he says, "I made up my mind I wouldn't try for it."

Despite its protests to the contrary, modern Christianity has become willy-nilly the religion of the state and the economic status quo. Because it has been so exclusively dedicated to incanting anemic souls into Heaven, it has been made the tool of much earthly villainy. It has, for the most part, stood silently by while a predatory economy has ravaged the world, destroyed its natural beauty and health, divided and plundered its human communities and households. It has flown the flag and chanted the slogans of empire. It has assumed with the economists that "economic forces" automatically work for good and has assumed with the industrialists and militarists that technology determines history. It has assumed with almost everybody that "progress" is good, that it is good to be modern and up with the times. It has admired Caesar and comforted him in his depredations and defaults. But in its de facto alliance with Caesar, Christianity connives directly in the murder of Creation. For in these days, Caesar is no longer a mere destroyer of armies, cities, and nations. He is a contradicter of the fundamental miracle of life. A part of the normal practice of his power is his willingness to destroy the world. He prays, he says, and churches everywhere compliantly pray with him. But he is praying to a God whose works he is prepared at any moment to destroy. What could be more wicked than that, or more mad?

The religion of the Bible, on the contrary, is a religion of the state and the status quo only in brief moments. In practice, it is a religion for the correction equally of people and of kings. And Christ's life, from the manger to the cross, was an affront to the established powers of his time, just as it is to the established powers of our time. Much is made in churches of the "good news" of the Gospels. Less is said of the Gospels' bad news, which is that Jesus would have been horrified by just about every "Christian" government the world has ever seen. He would be horrified by our government and its works, and it would be horrified by him. Surely no sane and thoughtful person can imagine any government of our time sitting comfortably at the feet of Jesus while he is saying, "Love your enemies, bless them that curse you, do good to them that hate you, and pray for them that despitefully use you and persecute you."

In fact, we know that one of the businesses of governments, "Christian" or not, has been to reenact the crucifixion. It has happened again and again and again. In *A Time for Trumpets*, his history of the Battle of the Bulge, Charles B.

MacDonald tells how the SS Colonel Joachim Peiper was forced to withdraw from a bombarded château near the town of La Gleize, leaving behind a number of severely wounded soldiers of both armies. "Also left behind," MacDonald wrote, "on a whitewashed wall of one of the rooms in the basement was a charcoal drawing of Christ, thorns on his head, tears on his cheeks — whether drawn by a German or an American nobody would ever know." This is not an image that belongs to history but rather one that judges it.

1992

The Pleasures of Eating

Many times, after I have finished a lecture on the decline of American farming and rural life, someone in the audience has asked, "What can city people do?"

"Eat responsibly," I have usually answered. Of course, I have tried to explain what I meant by that, but afterwards I have invariably felt that there was more to be said than I had been able to say. Now I would like to attempt a better explanation.

I begin with the proposition that eating is an agricultural act. Eating ends the annual drama of the food economy that begins with planting and birth. Most eaters, however, are no longer aware that this is true. They think of food as an agricultural product, perhaps, but they do not think of themselves as participants in agriculture. They think of themselves as "consumers." If they think beyond that, they recognize that they are passive consumers. They buy what they want—or what they have been persuaded to want—within the limits of what they can get. They pay, mostly without protest, what they are charged. And they mostly ignore certain critical questions about the quality and the cost of what they are sold: How fresh is it? How pure or clean is it, how free of dangerous chemicals? How far was it transported, and what did transportation add to the cost? How much did manufacturing or packaging or advertising add to the cost? When the food product has been manufactured or "processed" or "precooked," how has that affected its quality or price or nutritional value?

Most urban shoppers would tell you that food is produced on farms. But most of them do not know what farms, or what kinds of farms, or where the farms are, or what knowledge or skills are involved in farming. They apparently have little doubt that farms will continue to produce, but they do not know how or over what obstacles. For them, then, food is pretty much an abstract idea—something they do not know or imagine—until it appears on the grocery shelf or on the table.

The specialization of production induces specialization of consumption. Patrons of the entertainment industry, for example, entertain themselves less and less and have become more and more passively dependent on commercial suppliers. This is certainly true also of patrons of the food industry, who have tended more and more to be *mere* consumers—passive, uncritical, and dependent. Indeed, this sort of consumption may be said to be one of the chief goals of industrial production. The food industrialists have by now persuaded millions of consumers to prefer food that is already prepared. They will grow, deliver, and cook your food for you and (just like your mother) beg you to eat it. That they do not yet offer to insert it, prechewed, into your mouth is only because they have found no profitable way to do so. We may rest assured that they would be glad to find such a way. The ideal industrial food consumer would be strapped to a table with a tube running from the food factory directly into his or her stomach.

Perhaps I exaggerate, but not by much. The industrial eater is, in fact, one who does not know that eating is an agricultural act, who no longer knows or imagines the connections between eating and the land, and who is therefore necessarily passive and uncritical—in short, a victim. When food, in the minds of eaters, is no longer associated with farming and with the land, then the eaters are suffering a kind of cultural amnesia that is misleading and dangerous. The current version of the "dream home" of the future involves "effortless" shopping from a list of available goods on a television monitor and heating precooked food by remote control. Of course, this implies and depends on a perfect ignorance of the history of the food that is consumed. It requires that the citizenry should give up their hereditary and sensible aversion to buying a pig in a poke. It wishes to make the selling of pigs in pokes an honorable and glamorous activity. The dreamer in this dream home will perforce know nothing about the kind or quality of this food, or where it came from, or how it was produced and prepared, or what ingredients, additives, and residues it contains—unless, that is, the dreamer undertakes a close and constant study of the food industry, in which case he or she might as well wake up and play an active and responsible part in the economy of food.

There is, then, a politics of food that, like any politics, involves our freedom. We still (sometimes) remember that we cannot be free if our minds and voices are controlled by someone else. But we have neglected to understand that we cannot be free if our food and its sources are controlled by someone else. The condition of the passive consumer of food is not a democratic condition. One reason to eat responsibly is to live free.

But if there is a food politics, there are also a food esthetics and a food ethics, neither of which is dissociated from politics. Like industrial sex, industrial eating has become a degraded, poor, and paltry thing. Our kitchens and other eating places more and more resemble filling stations, as our homes more and more resemble motels. "Life is not very interesting," we seem to have decided. "Let its satisfactions be minimal, perfunctory, and fast." We hurry through our meals to go to work and hurry through our work in order to "recreate" ourselves in the evenings and on weekends and vacations. And then we hurry, with the greatest possible speed and noise and violence, through our recreation—for what? To eat the billionth hamburger at some fast-food joint hell-bent on increasing the "quality" of our life? And all this is carried out in a remarkable obliviousness to the causes and effects, the possibilities and the purposes, of the life of the body in this world.

One will find this obliviousness represented in virgin purity in the advertisements of the food industry, in which food wears as much makeup as the actors. If one gained one's whole knowledge of food from these advertisements (as some presumably do), one would not know that the various edibles were ever living creatures, or that they all come from the soil, or that they were produced by work. The passive American consumer, sitting down to a meal of pre-prepared or fast food, confronts a platter covered with inert, anonymous substances that have been processed, dyed, breaded, sauced, gravied, ground, pulped, strained, blended, prettified, and sanitized beyond resemblance to any part of any creature that ever lived. The products of nature and agriculture have been made, to all appearances, the products of industry. Both eater and eaten are thus in exile from biological reality. And the result is a kind of solitude, unprecedented in human experience, in which the eater may think of eating as, first, a purely commercial transaction between him and a supplier and then as a purely appetitive transaction between him and his food.

And this peculiar specialization of the act of eating is, again, of obvious benefit to the food industry, which has good reasons to obscure the connection between food and farming. It would not do for the consumer to know that the hamburger she is eating came from a steer who spent much of his life standing deep in his own excrement in a feedlot, helping to pollute the local streams, or that the calf that yielded the veal cutlet on her plate spent its life in a box in which it did not have room to turn around. And, though her sympathy for the slaw might be less tender, she should not be encouraged to meditate on the hygienic and biological implications of mile-square fields of cabbage, for vegetables grown in huge monocultures are dependent on toxic chemicals—just as animals in close confinement are dependent on antibiotics and other drugs.

The consumer, that is to say, must be kept from discovering that, in the food industry—as in any other industry—the overriding concerns are not quality and health, but volume and price. For decades now the entire industrial food economy, from the large farms and feedlots to the chains of supermarkets and fast-food restaurants, has been obsessed with volume. It has relentlessly increased scale in order to increase volume in order (presumably) to reduce costs. But as scale increases, diversity declines; as diversity declines, so does health; as health declines, the dependence on drugs and chemicals necessarily increases. As capital replaces labor, it does so by substituting machines, drugs, and chemicals for human workers and for the natural health and fertility of the soil. The food is produced by any means or any shortcut that will increase profits. And the business of the cosmeticians of advertising is to persuade the consumer that food so produced is good, tasty, healthful, and a guarantee of marital fidelity and long life.

It is possible, then, to be liberated from the husbandry and wifery of the old household food economy. But one can be thus liberated only by entering a trap (unless one sees ignorance and helplessness as the signs of privilege, as many people apparently do). The trap is the ideal of industrialism: a walled city surrounded by valves that let merchandise in but no consciousness out. How does one escape this trap? Only voluntarily, the same way that one went in: by restoring one's consciousness of what is involved in eating; by reclaiming responsibility for one's own part in the food economy. One might begin with the illuminating principle of Sir Albert Howard's *The Soil and Health*, that we should understand "the whole problem of health in soil, plant, animal, and man as one great subject." Eaters, that is, must understand that eating takes place inescapably in the world, that it is inescapably an agricultural act, and that how we eat determines, to a considerable extent, how the world is used.

This is a simple way of describing a relationship that is inexpressibly complex. To eat responsibly is to understand and enact, so far as one can, this complex relationship. What can one do? Here is a list, probably not definitive:

1. Participate in food production to the extent that you can. If you have a yard or even just a porch box or a pot in a sunny window, grow something to eat in it. Make a little compost of your kitchen scraps and use it for fertilizer. Only by growing some food for yourself can you become acquainted with the beautiful energy cycle that revolves from soil to seed to flower to fruit to food to offal to decay, and around again. You will be fully responsible for any food that you grow for yourself, and you will know all about it. You will appreciate it fully, having known it all its life.
2. Prepare your own food. This means reviving in your own mind and life the arts of kitchen and household. This should enable you to eat more cheaply, and it will give you a measure of "quality control": you will have some reliable knowledge of what has been added to the food you eat.
3. Learn the origins of the food you buy, and buy the food that is produced closest to your home. The idea that every locality should be, as much as possible, the source of its own food makes several kinds of sense. The locally produced food supply is the most secure, the freshest, and the easiest for local consumers to know about and to influence.
4. Whenever possible, deal directly with a local farmer, gardener, or orchardist. All the reasons listed for the previous suggestion apply here. In addition, by such dealing you eliminate the whole pack of merchants, transporters, processors, packagers, and advertisers who thrive at the expense of both producers and consumers.
5. Learn, in self-defense, as much as you can of the economy and technology of industrial food production. What is added to food that is not food, and what do you pay for these additions?
6. Learn what is involved in the *best* farming and gardening.
7. Learn as much as you can, by direct observation and experience if possible, of the life histories of the food species.

The last suggestion seems particularly important to me. Many people are now as much estranged from the lives of domestic plants and animals (except for flowers and dogs and cats) as they are from the lives of the wild ones. This is regrettable, for these domestic creatures are in diverse ways attractive; there is much pleasure in knowing them. And farming, animal husbandry, horticulture, and gardening, at their best, are complex and comely arts; there is much pleasure in knowing them, too.

It follows that there is great *dis*pleasure in knowing about a food economy that degrades and abuses those arts and those plants and animals and the soil from which they come. For anyone who does know something of the modern history of food, eating away from home can be a chore. My own inclination is to eat seafood instead of red meat or poultry when I am traveling. Though I am by no means a vegetarian, I dislike the thought that some animal has been made miserable in order to feed me. If I am going to eat meat, I want it to be from an animal that has lived a pleasant, uncrowded life outdoors, on bountiful pasture, with good water nearby and trees for shade. And I am getting almost as fussy about food plants. I like to eat vegetables and fruits that I know have lived happily and healthily in good soil, not the products of the huge, bechemicaled factory-fields that I have seen, for example, in the Central Valley of California. The industrial farm is said to have been patterned on the factory production line. In practice, it looks more like a concentration camp.

The pleasure of eating should be an *extensive* pleasure, not that of the mere gourmet. People who know the garden in which their vegetables have grown and know that the garden is healthy will remember the beauty of the growing plants, perhaps in the dewy first light of morning when gardens are at their best. Such a memory involves itself with the food and is one of the pleasures of eating. The knowledge of the good health of the garden relieves and frees and comforts the eater. The same goes for eating meat. The thought of the good pasture and of the calf contentedly grazing flavors the steak. Some, I know, will think it bloodthirsty or worse to eat a fellow creature you have known all its life. On the contrary, I think it means that you eat with understanding and with gratitude. A significant part of the pleasure of eating is in one's accurate consciousness of the lives and the world from which food comes. The pleasure of eating, then, may be the best available standard of our health. And this pleasure, I think, is pretty fully available to the urban consumer who will make the necessary effort.

I mentioned earlier the politics, esthetics, and ethics of food. But to speak of the pleasure of eating is to go beyond those categories. Eating with the fullest pleasure—pleasure, that is, that does not depend on ignorance—is perhaps the profoundest enactment of our connection with the world. In this pleasure we experience and celebrate our dependence and our gratitude, for we are living from mystery, from creatures we did not make and powers we cannot comprehend. When I think of the meaning of food, I always remember these lines by the poet William Carlos Williams, which seem to me merely honest:

There is nothing to eat,
    seek it where you will,
        but of the body of the Lord.
The blessed plants
    and the sea, yield it
        to the imagination
intact.

1989

Acknowledgments

Several of Wendell Berry's essays have appeared in multiple texts. The following bibliographic information is therefore not exhaustive. It lists where each essay is readily accessible.

"A Native Hill"—Wendell Berry. *Recollected Essays 1965–1980* (New York: North Point Press, 1981, pp. 73–113)

"The Unsettling of America"— Wendell Berry. *The Unsettling of America: Culture & Agriculture* (San Francisco: Sierra Club Books, 3rd Edition, 1996, pp. 3–14)

"Racism and the Economy"—Wendell Berry. *The Hidden Wound* (The Afterword, New York: North Point Press, 1989, pp. 112–137)

"Feminism, the Body, and the Machine"—Wendell Berry. *What Are People For?* (New York: North Point Press, 1990, pp. 178–196)

"Think Little"—Wendell Berry. *A Continuous Harmony* (New York: Harcourt Brace & Company, 1972, 1971, 1970, pp. 71–85)

"The Body and the Earth"—(*The Unsettling of America*, pp. 97– 140)

"Men and Women in Search of Common Ground"—Wendell Berry. *Home Economics* (New York: North Point Press, 1987, pp. 112–122)

"Health Is Membership"—Wendell Berry. *Another Turn of the Crank* (Washington, D.C.: Counterpoint Press, 1995, pp. 86–109)

"Sex, Economy, Freedom, and Community" (Abridged)—Wendell Berry. *Sex, Economy, Freedom & Community* (New York: Pantheon Books, 1992, 1993, pp. 125–131, 144–173)

"People, Land, and Community"—Wendell Berry. *Standing By Words* (San Francisco: North Point Press, 1983, pp. 64–79)

"Conservation and Local Economy" (*Sex, Economy, Freedom & Community*, pp. 3–18)

"Economy and Pleasure" (*What Are People For?* pp. 129–144)

"Two Economies" (*Home Economics*, pp. 54–75)

"The Whole Horse" (A shorter version of this essay appeared in *The Land Report* #64, Summer 1999, pp. 3–7)

"The Idea of a Local Economy" (*Orion* 20:1, Winter, 2001, pp. 28–37)
"A Bad Big Idea" (*Sex, Economy, Freedom & Community*, pp. 45–51)
"Solving for Pattern"—Wendell Berry. *The Gift of Good Land: Further Essays Cultural and Agricultural* (New York: North Point Press, 1981, pp. 134–145)
"The Use of Energy" (*The Unsettling of America*, pp. 81–95)
"The Gift of Good Land" (*The Gift of Good Land*, pp. 267–281)
"Christianity and the Survival of Creation" (*Sex, Economy, Freedom & Community*, pp. 93–116)
"The Pleasures of Eating" (*What Are People For?* pp. 145–152)

The author and editor wish to thank the following publishers: From *What Are People For?* (New York: North Point Press, 1989); *Home Economics* (New York: North Point Press, 1987); *Standing By Words* (New York: North Point Press, 1983); *Gift of Good Land* (New York: North Point Press, 1981), © Wendell Berry. Reprinted by permission of North Point Press, a division of Farrar, Straus & Giroux. Three chapters of this book, "The Unsettling of America," "The Use of Energy," and "The Body and the Earth" are from *The Unsettling of America: Culture & Agriculture*, © 1977 by Wendell Berry. All rights reserved. Reprinted by permission of Sierra Club Books. From *Sex, Economy, Freedom and Community* (New York: Pantheon Books, 1992), © Wendell Berry. Reprinted by permission.